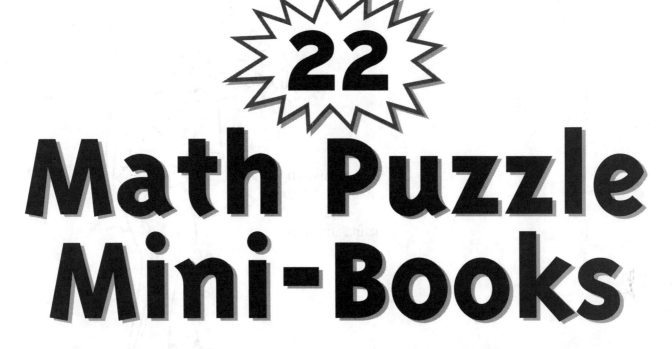

22
Math Puzzle
Mini-Books

Reproducible Mini-Books Filled With Brain-Tickling Math Puzzles for Your Students to Complete

By Michael Schiro & Rainy Cotti

SCHOLASTIC
PROFESSIONAL BOOKS

New York ◎ Toronto ◎ London ◎ Auckland ◎ Sydney ◎ Mexico City ◎ New Delhi ◎ Hong Kong

To our children
Stephanie and Arthur
and
Shanna and Kevin

Cover design by Pamela Simmons and Jaime Lucero

Interior design by Ellen Matlach Hassell
for Boultinghouse & Boultinghouse, Inc.

Interior illustrations by Teresa Anderko and Manuel Rivera

ISBN 0-590-91809-5

Contents

MINI-BOOKS

About This Book

This book features 22 mini-books, each of which contains seven mathematical problems of a similar type that are fun to solve. The seven problems in each mini-book are arranged in order of difficulty to lead students to an understanding of one class of mathematical problems. The problems come from such diverse fields as algebra, geometry, game theory, and topology—topics usually not accessible to students. In this way, they gain insight into the wide range of topics that mathematicians explore—and discover that mathematics is fun.

The mini-books are designed to overcome one of the difficulties with traditional presentations of nonroutine mathematical problems—that problems are presented as isolated entities for students to solve. As a result, students engaging in problem solving usually do not gain an understanding of overarching mathematical principles that describe the behavior of a whole class of problems. They usually do not have a chance to look back at problems just completed in order to formulate and test new mathematical hypotheses on successive problems of a similar type. *22 Math Puzzle Mini-Books* rectifies this by organizing each mini-book as a collection of seven carefully sequenced problems in a single field of mathematics that lead students to generate hypotheses about a whole class of problems.

Mathematical Problem Solving

Mathematical problem solving involves understanding the problem to be solved, devising a plan to solve it, carrying out the plan, and looking back to see the effectiveness of the plan (Polya 1957). These steps frequently overlap and are repeated as plans fail, which allows students to look back at a problem and understand it in such a way that a new plan can be devised.

Understand the Problem. Understanding a problem, puzzle, game, or strategy involves such things as being able to clearly describe it, delineate its parts and their relationships to each other, and understand the relationships among any of the problem's unknowns, data, conditions, or limitations.

Devise a Plan. Devising a plan or strategy for solving a problem grows out of repeatedly trying to solve a problem, noticing patterns, making hypotheses about those patterns, and testing the hypotheses.

Carry out the Plan. In carrying out a plan, a student must use the strategy devised. In doing so, students should check to see if their plans have the desired effect, if a new perspective on the problem is needed, or if an easier or more elegant way of solving the problem might be discovered. Evaluating and revising a plan is an important part of carrying it out.

Look Back. Once a strategy for solving a problem has been tested, it is important to look back to see if the strategy was successful and if it holds clues to other problems confronted and not yet solved. Reflecting on what has been accomplished often provides a metacognitive perspective on a problem and its relationship to other problems, greater understanding by consolidating newly constructed knowledge, and advance organizers useful in approaching similar problems. To facilitate looking back, encourage students to discuss and write about their understanding of problems and solutions.

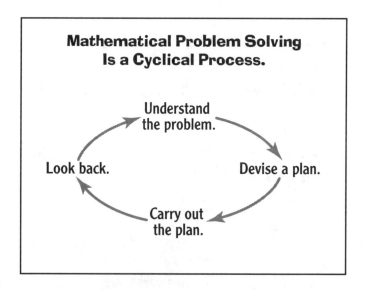

Mathematical Problem-Solving Strategies

Some strategies that can aid students in devising plans for solving problems are listed below.

- ◎ brainstorm
- ◎ act out problem
- ◎ look for patterns
- ◎ make a diagram
- ◎ write an equation
- ◎ guess and check
- ◎ make a model
- ◎ make an organized list
- ◎ make a table
- ◎ logical reasoning
- ◎ draw a picture
- ◎ solve a simpler problem
- ◎ work backward
- ◎ use objects to simulate a problem
- ◎ examine a similar, familiar problem

The traditional approach to teaching these strategies is to describe them at a theoretical level and then have students practice them on problems. This book suggests another approach, for students seem naturally to use almost all of the problem-solving strategies if they are simply allowed to solve a wide range of rich mathematical problems. Here is what we suggest:

1. Have students work on the problems in a mini-book. While this is occurring, carefully observe which strategies they use.

2. After students have solved a problem, discuss as a class how they solved the problem. As students describe their strategies, elaborate on each strategy by reference to what others in the class have done, name the strategies, and display a list of the strategies for future reference.

3. As more problems are solved, students should identify the strategies they use, add new strategies to the list, and further elaborate on what each strategy means—in light of their personal experiences.

In general, help students observe and learn problem solving strategies they and their peers use naturally. If this approach is taken, students seem to feel greater ownership of strategies; they seem to better comprehend the strategies; and they seem to better understand that many strategies exist, that there is no one best strategy, and that there are usually different ways to solve problems.

Problem Solving, Reasoning, Communication, Connections

In *Curriculum and Evaluation Standards for School Mathematics* (1989), the National Council of Teachers of Mathematics attempts to broaden the view of school mathematics to include problem solving, reasoning, communication, and connections. This book supports those NCTM initiatives.

- ◎ Mathematics as problem solving is encountered when students search for solutions to problems, puzzles, and game strategies as well as through their endeavors to discover the relationship among problems in each mini-book.

- ◎ Mathematics as reasoning is encouraged when students analyze problems, generate and test hypotheses, justify the usefulness of hypotheses, and build models that describe how to understand and solve a set of similar problems.

- ◎ Mathematics as communication is emphasized when students clarify their ideas during discussions or in writing as they offer conjectures and arguments to help them and their peers understand problems and their solutions.

- ◎ Mathematical connections are stressed when students see connections between branches of mathematics, use ideas of one branch to understand problems in another, and see how mathematics can help them comprehend their world.

How to Use the Mini-Books in the Classroom

Cooperative Groups The games in this book are usually for two people. The problems and puzzles are usually for one person. Students learn most about games, problems, and puzzles when they work in cooperative groups and discuss evolving strategies.

It is suggested that, for much of the work, students work in cooperative groups of from two to four persons. The goal of the groups should be to help every member discover solutions for a mini-book. The goal is *not* for groups to compete against each other. Groups should also share information with each other so that everyone in class understands the mathematics. To encourage this, whole-class discussions should be held periodically to share discoveries. Doing so raises two issues.

First, how can cooperative work be supported on such things as magic tricks or strategy games since once students learn tricks or strategies, they will want to use them on peers? Encourage students to work together and show tricks or play games with friends in different classes or with family members. Urge them to view classmates as colleagues who help devise strategies to use on others.

Second, how can you make sure that the student who first discovers a solution does not tell it to everyone and spoil their ability to uncover the solution themselves? We tell students that they can give a hint to help someone discover something, but they can never tell the answer outright and deprive someone of the fun of making a discovery. Small-group activity directed toward making discoveries must be distinguished from whole-group discussion directed toward sharing meaning.

Two practices are useful in facilitating groups. First, structure groups so that members see themselves as linked together in such a way that no one succeeds unless all members succeed. If the group as a whole succeeds, then all its members have succeeded. Second, have students help each other learn by being peer tutors and by providing constructive feedback.

Classroom Communication Communication helps students construct mathematical meanings. Math talk, math writing, math drawing, math gesturing, and math demonstrations with manipulatives can all help them make discoveries, construct meanings, and build understanding. Students should communicate with their peers and you the teacher as they work on the mini-books.

If students write "memos of understanding, explanation, or proof" that explain their understanding of problems and solutions, this helps them construct mathematical meanings. For instance, one teacher asks students to answer the following questions on finishing some mini-books:

◎ *Which were your most and least favorite problems? Explain your choices.*

◎ *If you were to rewrite the mini-book, what would you change?*

◎ *If you could tell the authors your thoughts about the mini-book, what would you say?*

Parents Inform parents about why you use mini-books. It can allay doubts that may arise when students go home and describe how they played with games, puzzles, or magic tricks during math class. Text for a sample note to parents is provided below.

Over the next several months, we will be working on problem solving during mathematics. Your child will be learning how to understand a mathematical problem, plan a solution to it, carry out the plan, and then look back to determine the success of the solution and its relationship to other mathematical problems. The problems your child will be working on come from arithmetic, geometry, algebra, and number theory.

The problems appear in mini-books, each of which contains seven problems in the form of puzzles, games, and even tricks. If your child comes home and wants to play a math game with you, let him or her teach you the game. If your child wants to play a math magic trick on you, have fun with the trick. Your child will learn how to create and solve problems of him or her own. You may see some of these as children learn to apply the skills and understandings they gain from engaging in problem solving.

You might also host a family math night where parents and siblings can work problems, solve puzzles, and play games with students. Set up four to six different learning stations, each centering on a mini-book. Students can run each station and organize family members as they perform magic tricks or play games. Later students can teach the families the underlying math. End the night with a talk about your math program. A family math night helps parents get to know your math program and gives them a sense of what can be learned from activities that are not textbook based.

Children Creating Their Own Problems
Mathematicians love to create new problems to solve. You may notice that the four-phase model of problem solving previously described lacks a first phase called "find a problem." In the context of the mini-books, finding a problem involves creating new problems, puzzles, or games that have the same underlying mathematical structure as those within each mini-book. Students should be supported in their efforts to create their own problems.

Comments on the Mathematics of Each Mini-Book

Bridge Crossing Problems

The branch of mathematics involved here is called network theory and was invented by Leonard Euler in the eighteenth century largely as a result of solving a real-life bridge crossing problem in Koenigsberg, Germany. The city had seven bridges and looked like the problem on page 4 of *MORE Bridge-Crossing Problems*. On Sundays, after church, the people of Koenigsberg would stroll around the city trying to walk across each bridge once and only once. They could never do it. Mathematicians from throughout Europe traveled to the city to try to solve the problem. How did Euler solve the problem? He solved the problem by showing that it was impossible to do.

Euler proved that the only bridge-crossing problems that can be solved are those with two or fewer landmasses, each containing an odd number of bridges. The number of landmasses with an even number of bridges makes no difference. Euler's proof went something like this:

⊚ *If a bridge-crossing problem has 1 landmass with an odd number of bridges, you must start or end your travels on that landmass.*

⊚ *If a problem has 2 landmasses, each with an odd number of bridges, you must start your travels on 1 of these landmasses and end it on the other landmass.*

⊚ *If a bridge-crossing problem has 3 or more landmasses, each with an odd number of bridges, you must either start or end your travels on each landmass—but this is impossible because you can start on only 1 landmass and end on another landmass.*

The bridge-crossing problems help students discover this proof. In the first mini-book, they should dis-

cover on which landmass to start and end in order successfully to complete the problem. **In the second mini-book, students should realize that if a problem has 3 or more landmasses, each with an odd number of bridges, then it is impossible to complete the problem.** Emphasize that the bridges must be walked or driven across.

Tracing Puzzles

Tracings are drawn with a pencil, without lifting the point from the page. No line segment may be traced more than once. Vertices (corners) may be passed through as many times as desired.

Some figures in the books cannot be traced. Figures can be traced only if they have 0, 1, or 2 odd vertices (an odd vertex is one that has an odd number of line segments attached to it). This is because if a tracing has odd vertices you must either start or end at an odd vertex. Since you can start at only one vertex and end at one vertex, only tracings with 2 or fewer odd vertices can be traced.

These tracing problems are identical (or isomorphic) to bridge-crossing problems. The vertices are the same as the landmasses in bridge-crossing problems, while the line segments are the same as bridges. That problems can look different and yet have the same underlying mathematical structure is one of the important ideas of mathematics.

To help students discover the relationship between tracing and bridge-crossing problems, have them work on tracing problems at least one week, but not more than one month, after completing the bridge problems. When students notice similarities between the problems, encourage them to describe the relationships, their tests for possible versus impossible problems, and their solution strategies.

Visual Estimation

This book is about visual illusions—things we perceive with our eyes that don't correspond to what is really in our world. Visual illusions have existed for thousands of years, yet they still baffle us. We still don't know exactly why many of the illusions exist. And the physics of explaining them is frequently quite complex. The purpose of this book is not to explain the illusions but to inquire how mathematics can help us distinguish illusion from reality.

The content of the book is categorized as visual estimation, for the intent is to have students first examine each illusion to make an estimation about "what appears to be" and then use mathematics to determine "what really is." To decide "what really is," students choose math tools—**other than rulers**—that might be helpful in determining the nature of reality and to check hypotheses. For instance, a simple caliper can be devised by making two marks on the side of a sheet of paper for one length and then moving the paper to check that length elsewhere. A straightedge exists on the edge of every sheet of paper.

Students should reflect on, discuss, and attempt to explain the saying on each page. Tell them that their explanations can reference examples from their own lives.

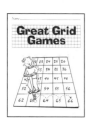

Number Grid Puzzles

Normally students learn about only very specific numerical relationships; for example, 24 + 9 = 33. With the hundreds chart, they see patterns that operate over all numbers simultaneously. For instance, students discover that adding 9 to any number results in a number one row down and one column to the left—or that it results in increasing the digits in the tens place by 1 and decreasing the digit in the ones place by 1

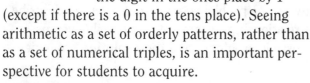

(except if there is a 0 in the tens place). Seeing arithmetic as a set of orderly patterns, rather than as a set of numerical triples, is an important perspective for students to acquire.

Many students wonder what happens if you fall off the hundreds chart; for example, what happens if you move to the right of 39 or below 95? Remind students that the hundreds chart does not stop at its edges—it keeps going. The hundreds chart is a snapshot of ten number lines stacked on top of each other; the number lines keep going. There are number lines for numbers in the hundreds and thousands that extend beyond the hundreds chart. The number lines even extend into negative numbers

The questions at the bottom of the mini-book pages focus attention on the mathematics underlying each puzzle. Students should discuss them in cooperative groups and write down their answers. The written answers can then be shared with the rest of the class to help students find language to explain what they have observed and reach higher levels of mathematical generalization.

Use *Great Grid Games* before *MORE Grid Games*, which is more difficult. Each puzzle in *MORE Grid Games* can be turned into a magic trick on the number grid that students can share with friends and parents. A reproducible for the hundreds chart appears in the Appendix on page 57. Duplicate and distribute it before students begin the mini-books so they can try out and record their ideas.

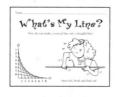

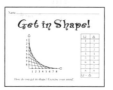

Line and Shape Designs

Line and shape designs combine algebra and geometry. Each design consists of an equivalent numerical table, an equation, and a geometrical construction. Understanding the equivalence that can exist between these different representations of a function is at the heart of algebra and analytic geometry.

As students work on the designs, ask them to explain how the table, the equation, and the geometric constructions relate to each other. After page 4 of *What's My Line?*, students must see the relationships in order to complete the designs.

The following question asked on page 8 of *What's My Line?* is crucial: *Why do straight lines make curves?* There is no one correct answer, but students should be able to describe how an "offset

saddle" creates the illusion of a curve. This way of making curves underlies much architecture—as can be seen in many buildings with curved arches or roofs.

Use *What's My Line?* before *Get in Shape!* The equations in *What's My Line?* are simpler. After students create the designs, consider asking them to complete the following map-coloring problem: *Color the spaces between the lines with two different color pencils in such a way so that no two spaces sharing an edge are the same color.* The curves created by most of the designs are parabolas. The curve in *Get in Shape!* on page 2 is a circle; on page 3 it is a cardioid.

You will find design frames on page 57 to duplicate so students can create more designs.

Using line designs in education is not new. Edith Somervell wrote *A Rhythmic Approach to Mathematics* about the subject in 1906. Teacher trainees at the beginning of the century created line designs with needle and thread by sewing designs on cardboard. Displaying students' work on a bulletin board will illustrate the effect of neatness, accuracy, imagination, and an understanding of geometry, particularly when they create their own designs.

Penny in Pocket Puzzles

These puzzles are a form of pre-algebra that are accessible to students without the need for algebraic notation or logical proof. What is necessary is a good understanding of what mathematical language means and what some of the implications of that language are, as well as the use of logical reasoning and the coordination of mathematical information found in separate statements.

Throughout the puzzles, words such as *difference*, *sum*, *most*, and *least* are used. Students will need to discuss the language in order to understand the problems. Discussion can stimulate a rich exchange of mathematical ideas and the clarification of meaning that they may not have fully constructed.

For instance, the puzzle on page 5 of *Pockets Full of Pennies* states the following: *There are pennies in two pockets. The sum of the number of pennies in the two pockets is 10. The difference in the number of pennies is 0.* This puzzle is easy to solve if you know how to translate each verbal statement into an algebraic equation: *Let x and y stand for the number of pennies in each pocket.* $x + y = 10$ and $x - y = 0$. *Solve for x and y.*

This, however, is not the objective of these puzzles. What students must determine is what it means when we say, "The difference in the number of pennies in the pockets is 0." To do so, they must explore such things as the following: *What does the word difference mean? Does it mean the same as subtract? If the difference is 0, is one number larger than the other number, or are they the same?*

Number rods can be helpful in solving some of these puzzles.

Nim Games

Most of the games have the same mathematical structure. "Dicey Count Up" is the same game as the others except players determine a winning number by throwing dice, which helps them generalize about winning strategies. What is important for students to discover is that things that look very different on the surface can be mathematically identical.

Underlying all of the nim games are winning numbers and a choice rule that tells you how many items you can count, pick up, subtract, color in, or move. In "Count Up," the player who says 21 loses, and by default, the player who says 20 wins, for the winning player forces the losing player to say 21. The number 20 is the first winning number that students will discover. To acquire 20, students will find that they must get 17, and that to get 17, they must say 14, that to get 14, they must acquire 11, and so on. The winning numbers for "Count Up" are 20, 17, 14, 11, 8, 5, and 2. The choice rule determines which numbers are winning numbers once the winning number is known. The winning numbers are 1 more apart than the size of the choice rule. In "Count Up," the choice rule is 2 since players can count only 1 or 2 numbers beyond where their opponents stopped counting, and thus the winning numbers are 3 numbers apart. Students

should realize that winning numbers are 3 numbers apart for the following reason: *If you say 17 and your opponent says 18, then you can say 19, 20 and get the last winning number of 20; if you say 17 and your opponent says 18, 19, then you can say 20.* Because of the size of the choice rule, no other choices are available other than saying 1 or 2 numbers, and thus the player who says 17 can get 20 and force his or her opponent to say 21.

Students should work in pairs. Their goal is to figure out and understand winning strategies. Strategies can be tested by playing the game many times with classmates and friends. When all students have found winning strategies, they should share their strategies in a large-group discussion.

Some students will uncover winning strategies quickly. Others will need more time. Give them time, and allow students to learn from and teach each other. Once a strategy for "Count Up" is discovered, strategies for the other games will quickly surface.

Mathematical Magic Tricks

These books introduce students to algebra. Underlying most of the tricks is a simple algebraic equation. For example, the equation for "Magic Sub 5" is $[4(x + 4) + 4] \div 4$, where x is the number selected by the magician's friend. The verbal representation is select a number, add 4 to it, multiply the sum by 4, add 4 to the product, divide that sum by 4. The magician subtracts 5 from the final number (the quotient) to find the number. That can be written as $[4(x + 4) + 4] \div 4 = x + 5$. When students explore why the tricks work, they are in essence exploring algebraic equations.

There is no need to use algebraic symbols or equations, although if you are comfortable using them, they can help clarify the tricks' magic. What is important is for students to explore how the tricks work and to find words to explain the "magic." Have them practice and explore the tricks in cooperative groups. Once students have mastered a trick, encourage them to try it on family and friends from other classes.

After students learn how a trick works, encourage them to create their own variations. For instance, in "Magic Sub 5," if the number 7 is used in all calculations, then the trick would be called "Magic Sub 8."

Here are some comments from a teacher who utilized these books in the classroom:

- When real-life magicians do tricks, they talk a lot to distract the audience from the essence of the trick. Encourage student magicians to "ham it up" with distracting statements. Learning to talk in front of an audience is important.

- Magicians do not repeatedly try the same trick on the same person, or else that person will figure out the trick. Suggest that students create a magic show that contains a series of tricks.

- The magic tricks provide students with a lot of arithmetic practice. **Be aware that the difficulty of the arithmetic increases as one progresses through the two books.**

- Try not to let students race through one trick after another. The tricks are to be savored, and discussions need to take place about why each trick works and how to clearly explain its mathematics. Doing one trick every other day may be enough.

- The inability to do the calculations can ruin a trick. In performing the trick in front of an audience, have the person who chooses a number share it with others so they can check all calculations.

- If students (or adults) have difficulty doing the arithmetic, they should use calculators to check their work.

Magic Circles and Magic Squares

About 2200 B.C., the Chinese discovered that a set of integers could be arranged in a square so that the sum of the integers in every row, column, and main diagonal is equal to the same constant sum. Such an array of numbers was believed to be magic. Legend has it that Emperor Yu discovered magic squares by studying the markings on the back of a turtle.

Knowledge of magic squares spread to India and Japan by the first century B.C. and to Europe about A.D. 1600.

Today we know that numbers can be arranged in many "magic shapes." It is also possible to use numbers other than integers. The following sets of numbers can be used in the magic cross (pages 2 and 3 of *Magic Squares*): [-2, -1, 0, +1, +2; sum of 0], [$\frac{1}{9}$, $\frac{2}{9}$, $\frac{3}{9}$, $\frac{4}{9}$, $\frac{5}{9}$; sum of 1], [0.25, 0.50, 0.75, 1, 1.25; sum of 2.25], [10%, 20%, 30%, 40%, 50%; sum of 90%].

Observe the number patterns. Use five consecutive numbers, each of which is larger than the last by a fixed amount, and the magic sum is the sum of the first, last, and middle numbers. With this observation, you and students should be able to invent many new puzzles.

Magic Squares includes two games, "Get 15" and "Hot." They are designed to show different aspects of magic shapes. "Get 15" can be played like tic-tac-toe if the integers from 1 to 9 are arranged in a 3-by-3 magic square summing up to 15. With this "magical secret" the game is easy to play. "Hot" works similarly. The words can be put in a magic square in such a way so that every row, column, and main diagonal has a word that contains the same letter. "Hot" shows that not just numbers can be put into magic shapes.

Crayon Digits and Crayon Constructions

These puzzle books use a familiar material, crayons, in an unfamiliar way. The puzzles may also be constructed out of other materials such as blunt toothpicks, coffee stirrers cut in half, or straws cut into small, equal-size pieces.

In addition to problem solving, these puzzles require the use of spatial and arithmetic concepts and skills. *Crazy Crayon Digits* uses numbers that have been constructed in the same style as the digits on digital watches and clocks, microwaves, and calculators. The structure of these numbers allows them to be transformed into other numbers by making minor changes. The first five

puzzles help students understand how the numbers are constructed. Then they apply this knowledge as they think about different ways of making false number sentences true. This also involves students in analyzing number relationships and whole-number operations.

The puzzles in *Creative Crayon Constructions* help develop students' spatial sense, too, as they analyze each construction, interpret what is being asked, and rearrange or remove crayons. After they complete these mini-books, students should be encouraged to create their own number sentence puzzles and crayon constructions.

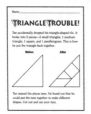

Tangrams

These tangram puzzle books help students discover attributes of geometric shapes and build their understanding of part-whole relationships. Reproducibles for the tangrams appear in the Appendix on page 57. Duplicate the tangrams for students and have them cut out the pieces to use with the puzzles.

Tangrams are said to originate in China. This fictional story extends the original legend of their creation. Share it with students before they begin *Triangle Trouble!* A Chinese man named Tan was hired by the emperor to make a special triangular tile to replace a broken one in the corner of a very old table. It took Tan many weeks to make the tile, using secret processes and clay found only in a remote part of China. While delivering the tile to the emperor, Tan tripped over a rock and dropped the triangular tile. It broke into five pieces. Tan sat down and wept. It would be weeks before he could make another tile. He tried to put the pieces back together, and as he did so, Tan noticed that it had broken into two small triangles, one medium triangle, one square, and one parallelogram. In reconstructing the tile, he also found that he could make many different shapes. But he realized that he still didn't have a tile for the emperor. Tan began to weep again. Just then a magic dragon flew down and landed at his feet. He told Tan not to be upset because the emperor loved games and puzzles and

would be thrilled to have such a puzzle tile in the table. It would amuse the emperor to make different shapes out of the pieces of the tile. Before Tan could say anything, the dragon disappeared. The dragon was right; the emperor loved the triangular tile, and he asked Tan to create a magic tile.

Coin Challenges

Coin Challenges contains puzzles and games for coins, but any round objects may be used, such as pennies, nickels, dimes, buttons, counters, and so on. Many of the problems can be worked on the puzzle pages, while others require more space. These puzzles will help students develop spatial sense and logical reasoning as well as some nonroutine thinking. All require perseverance.

Mystery Monsters

Mystery monsters puzzles motivate students to identify and discriminate characteristics of "monsters." They must focus on attributes, using logical reasoning while looking for similarities and differences in the monsters. Students must identify the characteristics, such as number of eyes or shape of monsters, that determine whether something does or does not belong to a particular class and then apply that knowledge to draw a "monster" that reflects those attributes.

The puzzles are sequenced in order of difficulty. Initially, one attribute defines a monster, then two, and finally three. In the first two puzzles, students decide if a creature belongs to a particular class of monsters. Remaining puzzles ask them to draw their own versions of monsters that exhibit the defining characteristics and to write how they know their version is one of the mystery monsters. Students are also invited to create their own mystery monster to share with friends.

Reference

G. Polya, *How to Solve It* (Princeton, NJ: Princeton University Press, 1957).

How to Make the Mini-Books

You'll Need: scissors, stapler

Directions for Photocopying:

1. Place the first page of the mini-book facedown on the platen glass, centering the dashed and solid lines in the copying area.

2. Make a photocopy for each student in the class.

3. Place the copies in the feeder tray with the blank side up. (Note: Some copiers may require that you place the copies with the blank side down.)

4. Turn the page over on the platen glass to copy the second page. Again, carefully center the dashed and solid lines in the copying area.

5. Make one copy to check that the pages are aligned correctly. The dashed and solid lines should match back to back, and the text should read in the same orientation. If they are aligned, make the remaining copies. (You may need to experiment a few times to discover the correct alignment.)

Directions for Assembly:

1. Have students cut apart the pages along the solid lines and then fold the pages along the dashed lines.

2. Help students assemble the pages in order, pointing out the page numbers in the lower corners. Slip pages 3–6 between pages 2 and 7.

3. Have them staple the book together along the folded edge.

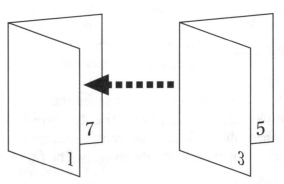

Create your own bridge crossing problem below.

I'll Cross That Bridge When I Get to It

This map shows a river with land on the outside, an island in the middle, and 4 bridges. The line shows how to cross each of the bridges once and only once, without swimming, jumping, boating, or flying across the river.

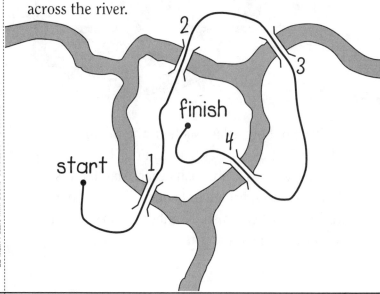

8

22 Math Puzzle Mini-Books Scholastic Professional Books

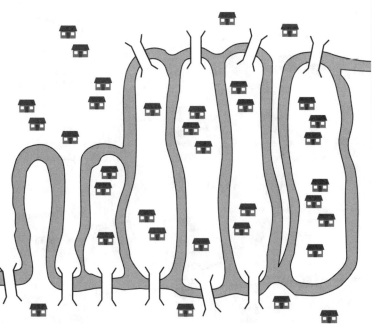

Mr. Chips delivers potato chips to all the islands in Munchy River. On his deliveries, he drives across each bridge only once. (No wonder Mr. Chips lives in his truck!) Show his route.

6

Juan is trapped inside a house on the middle island! Zelda must save him! She must cross each of the bridges over Zany River once and only once. Draw her route and number the bridges.

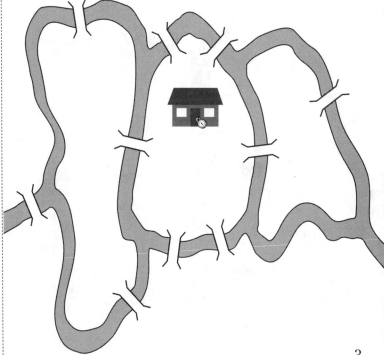

3

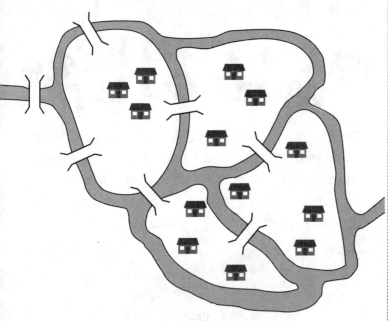

The village of Four Islands sits in the middle of the Red River. Wendy crosses over to Four Islands every day to deliver bread. She visits each island, but she walks across each bridge only once. Draw Wendy's route from start to finish. Number the bridges in the order that she crosses them.

2

Jared and Brooke just finished eating at Three Island Pizza Palace. Jared says it's impossible for them to walk across all of Three Island's bridges once and only once. Brooks says she can. Show her route.

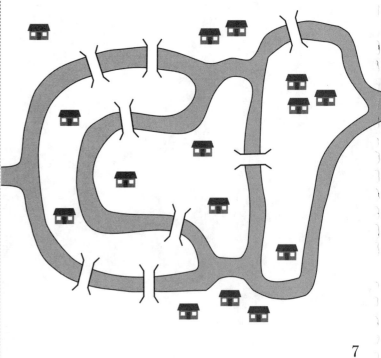

22 Math Puzzle Mini-Books Scholastic

7

Rose Ann drives the school bus. In order to pick up all the children in her town, she must cross many bridges—but only once! Draw her route and number the bridges.

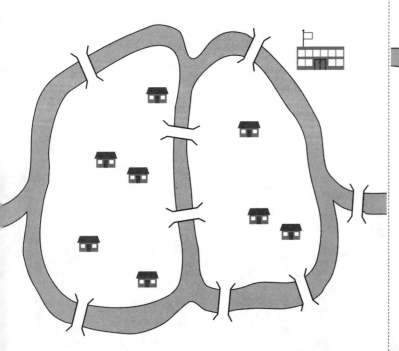

4

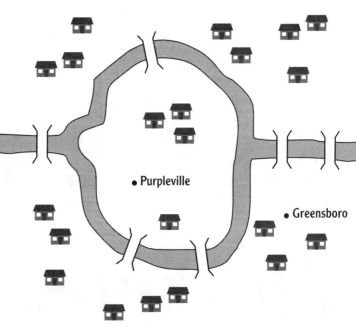

Karla delivers the mail. Karla's delivery route is between Purpleville and Greensboro. Show how she can cross each bridge once and only once.

5

Now it's your turn!
Invent, draw, and write your
own bridge-crossing problem.

Name_____

MORE Bridge-Crossing Problems

These bridge crossings are more difficult to solve.

Remember, you can cross each bridge only once.
On each puzzle, draw your route.

22 Math Puzzle Mini-Books Scholastic Professional Books

An ice cream store in Twelve Bridges is holding a contest. Whoever can cross each of the town's bridges once and only once to get to the store wins 12 free ice cream cones. Can you show a way? Where do you think the store is located? Draw it on the page.

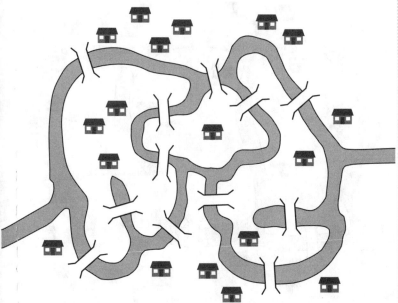

Flip Flop Jones owns two homes on the river. He wanders up and down and across and along the river selling apples. Show how Flip Flop can do this every day without crossing a bridge more than once. Where do you think his homes are located? Draw them on the page.

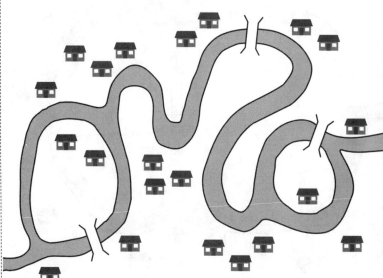

The town of Eyes on the Ball River has 10 bridges. After school, the children go to the park. They cross each bridge once and only once. Can you show their route? Where do you think the school and the park are located? Draw them on the page.

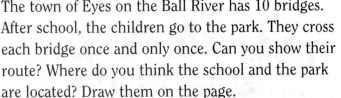

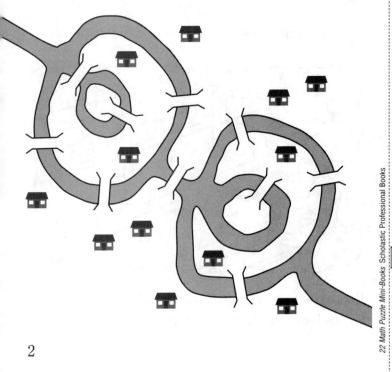

22 Math Puzzle Mini-Books Scholastic Professional Books

Roberto is lost on an island in the middle of the Palomino River. He wants to get off of the island and onto to solid ground. Can you lead Roberto off the island by crossing each bridge once and only once?

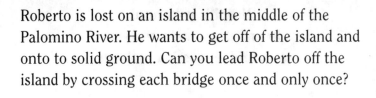

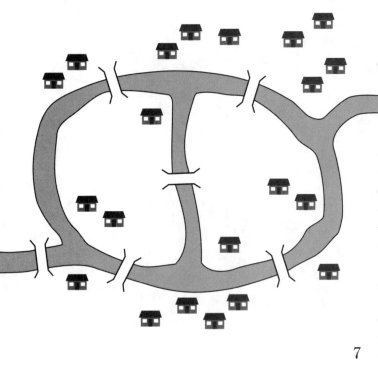

Every Sunday afternoon, Maria used to walk around her town and cross the bridges. Can you show how she might cross each bridge once and only once?

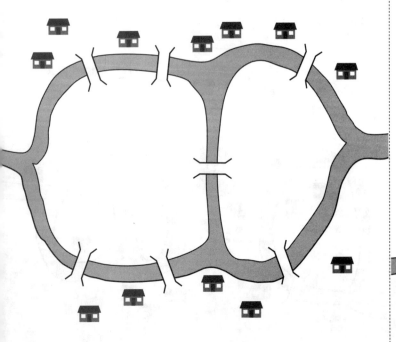

Walter collects money from the city's parking meters. On his route, he tries to cross each bridge over the Straight River only once. Can you draw how Walter can do this?

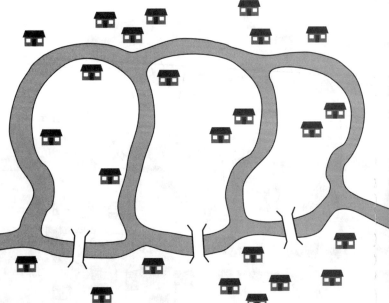

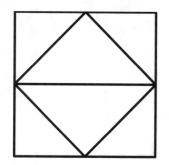

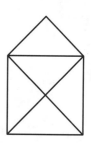

Draw each figure in the box without lifting your pencil point from the page. You may **not** go back over a line segment. You may go through corners more than once. Number the line segments in the order that you drew them.

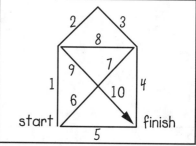

22 Math Puzzle Mini-Books Scholastic Professional Books

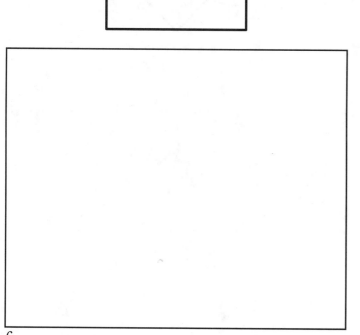

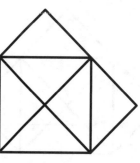

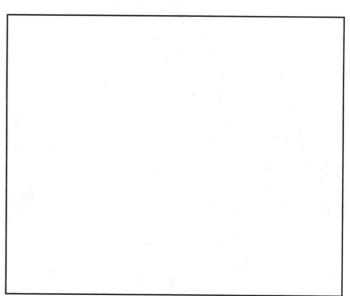

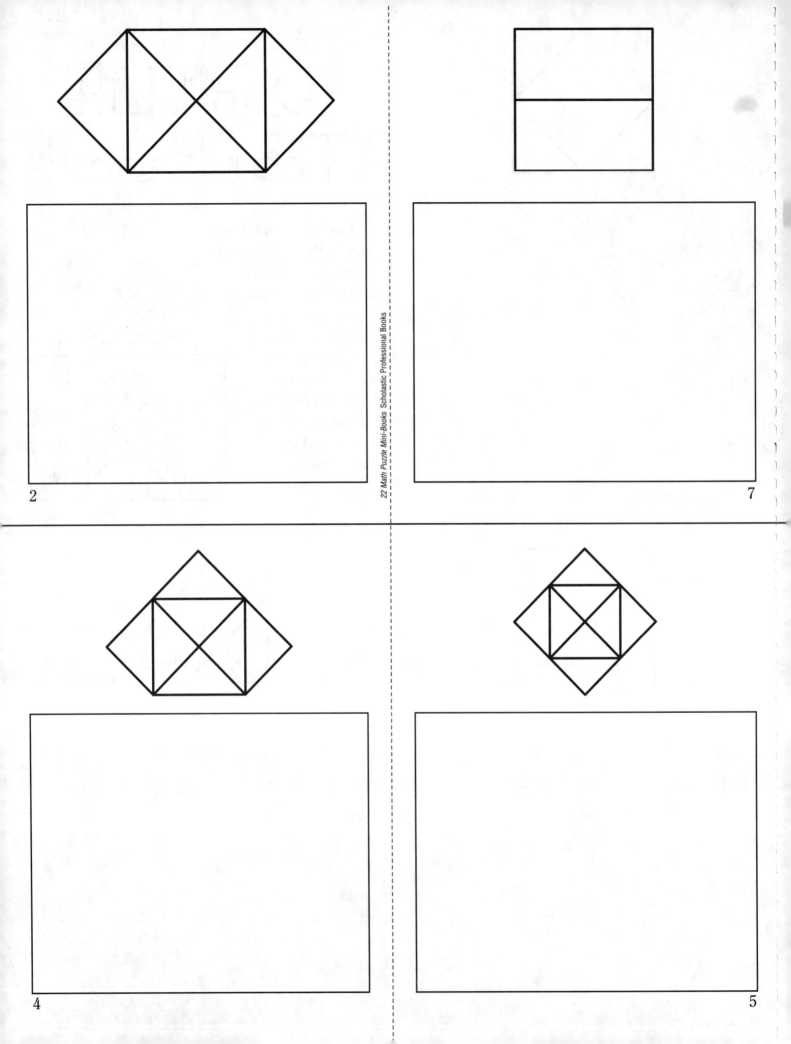

22 Math Puzzle Mini-Books Scholastic Professional Books

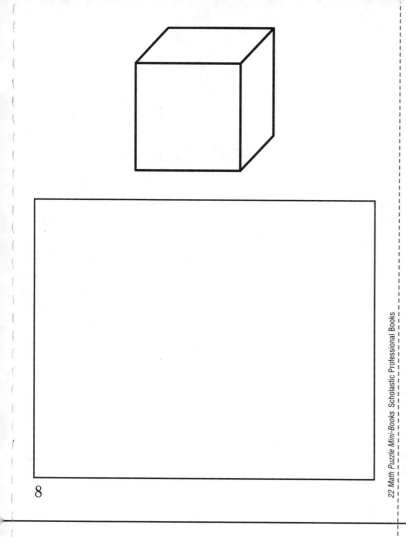

8

MORE
Tracing Puzzles

The puzzles in this book are trickier, but the rules are the same. Draw each figure without lifting your pencil point from the page. You may **not** go back over a line segment. You may go through corners more than once. Number the line segments in the order that you drew them.

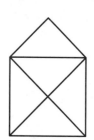

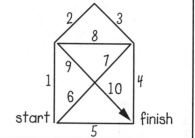

22 Math Puzzle Mini-Books Scholastic Professional Books

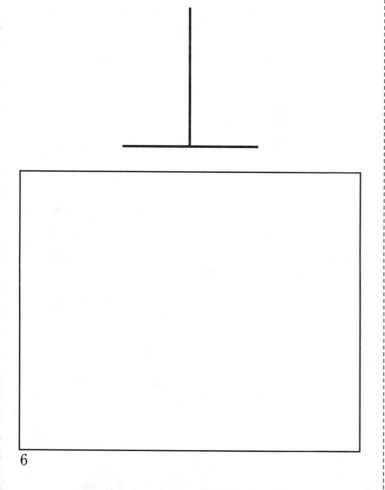

6

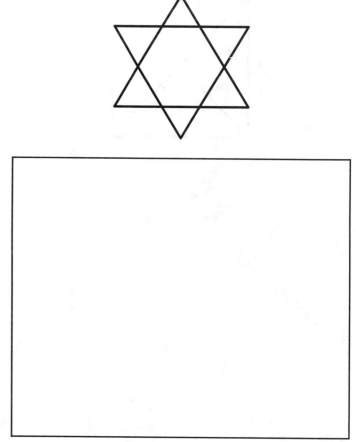

3

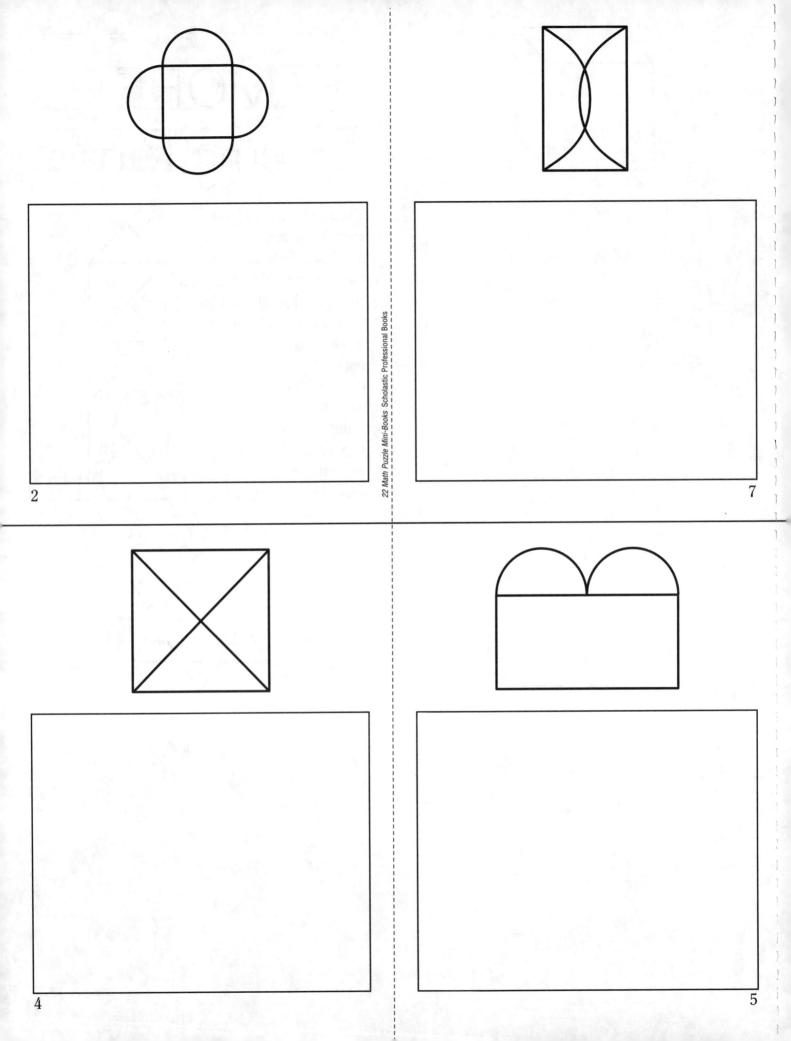

22 Math Puzzle Mini-Books Scholastic Professional Books

"I see and I believe.
I measure and I know."

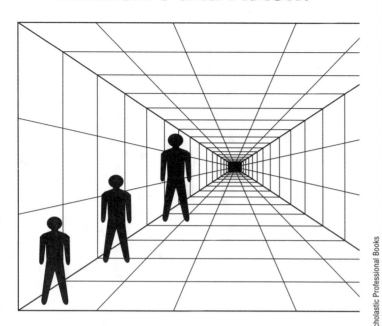

Are the people different sizes? **Estimate.**
Which math tools can you use to find out? **Try it.**

8

22 Math Puzzle Mini-Books Scholastic Professional Books

Don't Believe Your Eyes!

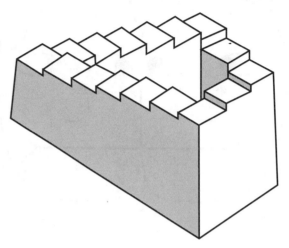

Which is the lowest step? **Estimate.**
Which math tools can help you find out? **Try it.**
Share all your answers with a friend. **Discuss it.**
And always wonder . . . **Why?**

"Not everything in mathematics is obvious."

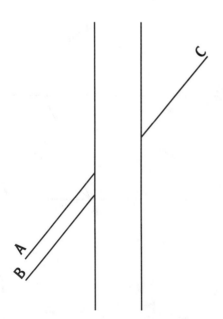

Will line A or line B extend to line C? **Estimate.**
Which math tools can you use to find out? **Try it.**

6

"Seeing is believing, or is it?"

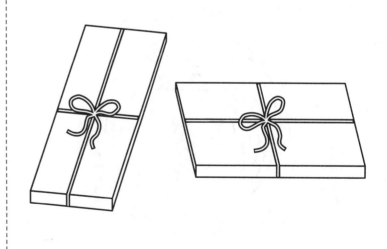

Which box top is larger? **Estimate.**
Which math tools can you use to find out? **Try it.**

3

"Not everything is as it seems."

Are the smile lines the same length? **Estimate.**
Which math tools can you use to find out? **Try it.**

2

22 Math Puzzle Mini-Books Scholastic Professional Books

"Let not your eyes be the only window to your world."

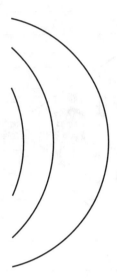

Which arc comes from the largest circle? **Estimate.**
Which math tools can you use to find out? **Try it.**

7

"The setting in which things are observed influences how you see them."

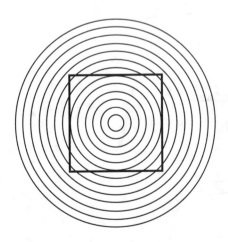

Are the sides of the square straight? **Estimate.**
Which math tools can you use to find out? **Try it.**

4

"Measure before you leap."

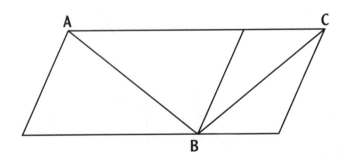

Is line *AB* the same length as line *BC*? **Estimate.**
Which math tools can you use to find out? **Try it.**

5

Operational Arrows

Create an arrow path for each problem below.
The first two problems have been done for you.
HINT: Each problem has more than one solution.

21 + [2] = 23 21 → → 23

21 + [11] = 32 21 ↓ ↑ ↓ → 32

35 + [ ] = 46 35 ________ 46

12 + [ ] = 45 12 ________ 45

38 + [ ] = 46 38 ________ 46

38 – [ ] = 26 38 ________ 26

55 – [ ] = 24 55 ________ 24

62 – [ ] = 30 62 ________ 30

Find five different paths from 62 to 30.
Which path is the shortest?

22 Math Puzzle Mini-Books Scholastic Professional Books

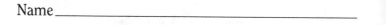

Name ________________

Great Grid Games

Use a hundreds chart with this book.

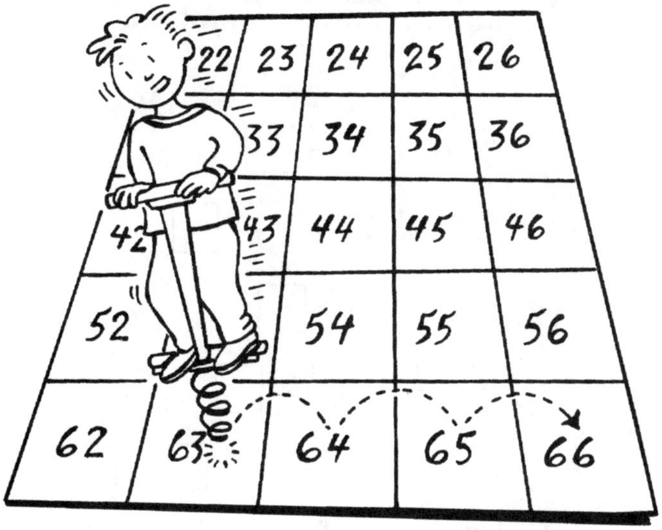

What's My Rule?

Look at the diagonal arrows on the grid below.
Try each movement on your number grid.
What calculation does each arrow do?

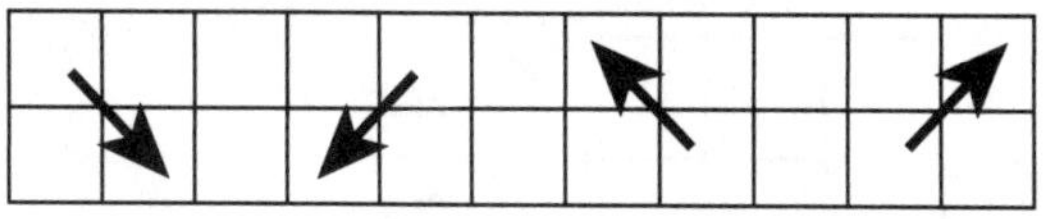

arrow		calculation
→	means	+ 1
←	means	– 1
↑	means	________
↓	means	________
↘	means	________
↖	means	________
↗	means	________
↙	means	________

Remember:

→ **Move 1 number to the right.**
← **Move 1 number to the left.**
↓ **Move 1 number down.**
↑ **Move 1 number up.**

(55) ↑ ↑ ↑ ↑ [ ]

(45) ↑ ↑ ↑ ← ← [ ]

(88) ↑ ↑ ↑ ↑ ← ← ← [ ]

(36) ← ← ← ← ← [ ]

(77) ← ← ↑ ↑ ↑ ↑ [ ]

(96) ← ← ↑ ↑ ↑ ↑ [ ]

How do these paths relate to subtraction? Why?

Arrow Paths

Start with the circled number on the number grid. Follow each arrow path. Write the number you end on in the square below.

→ Move 1 number to the right.
← Move 1 number to the left.
↓ Move 1 number down.
↑ Move 1 number up.

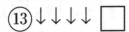

(13) ↓ ↓ ↓ ↓ □

(22) ↓ ↓ → → □

(55) ↓ ↓ ↓ → → → □

(33) → → → □

(65) → → ↓ ↓ ↓ □

(41) → → → → ↓ ↓ □

Tell how these paths relate to addition.

22 Math Puzzle Mini-Books Scholastic Professional Books

What's My Rule?

Each arrow describes a number pattern on the grid. Using the shaded squares, find three numbers on the grid to fit the pattern. Extend it for two more numbers. What's the rule?

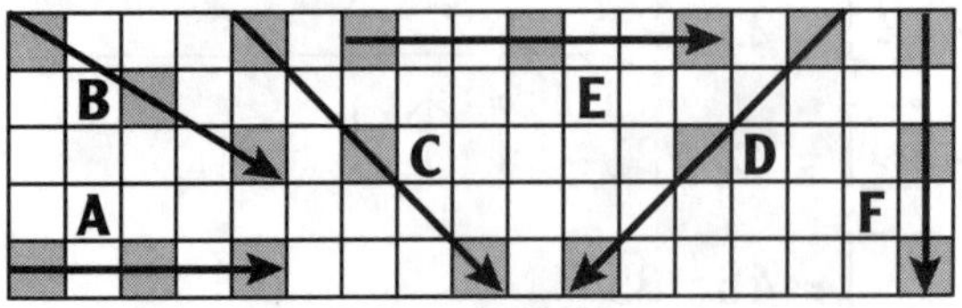

arrow	numbers					rule
A	20	22	24	26	28	+2
B						
C						
D						
E						
F						

Make up some of your own puzzles to share with friends.

Number Grid Puzzles

Complete each piece of the number grid. What number goes in the circle? Can you do it without looking at a number grid?

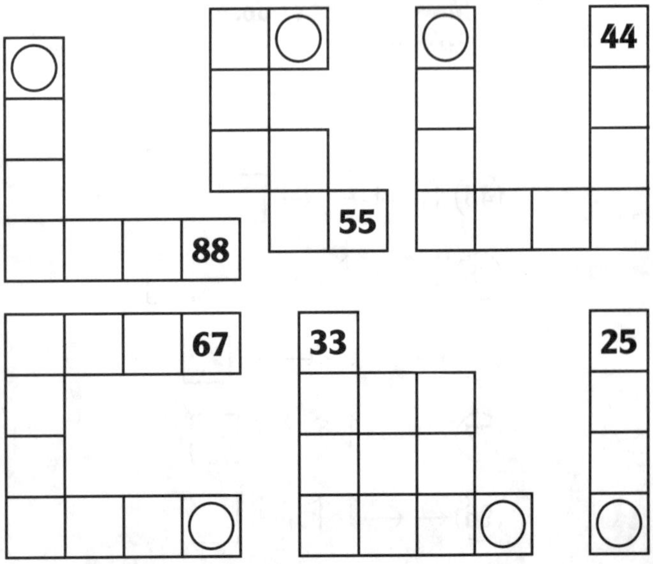

Can you find more than one way to do this?

More Grid Puzzles

What number goes in each circle? Try to find out without looking at the number grid. Can you find more than one way of solving the puzzles?

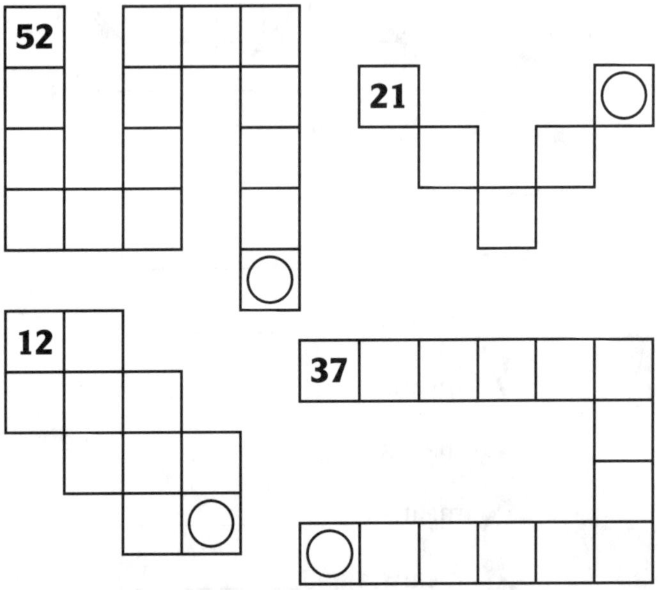

Make up some grid puzzles yourself. Share them with a friend.

Multiplying Opposite Corners of Four Small Squares

Find any four small squares that meet in a single corner. Multiply the numbers in the upper left and lower right corners. Multiply the numbers in the upper right and lower left corners. Compare products.

0	1	2	3	4	5	6	7	8	9
10	11	12	13	14	15	16	17	18	19
20	21	22	23	24	25	26	27	28	29
30	31	32	33	34	35	36	37	38	39
40	41	42	43	44	45	46	47	48	49
50	51	52	53	54	55	56	57	58	59
60	61	62	63	64	65	66	67	68	69
70	71	72	73	74	75	76	77	78	7

What happens with any 3-by-3 square? Explain why.
What happens with any 4-by-4 square? Explain why.

8

22 Math Puzzle Mini-Books Scholastic Professional Books

Name_____

Explore some numbers on a hundreds grid.

Discover a number pattern.

Check to see if the pattern works everywhere on the grid.

Describe the number pattern in words.

Explain in words why the number pattern works.

0	1	2	3	4	5	6	7	8	9
10	11	12	13	14	15	16	17	18	19
20	21	22	23	24	25	26	27	28	29
30	31	32	33	34	35	36	37	38	39
40	41	42	43	44	45	46	47	48	49
50	51	52	53	54	55	56	57	58	59
60	61	62	63	64	65	66	67	68	69
70	71	72	73	74	75	76	77	78	79
80	81	82	83	84	85	86	87	88	89
90	91	92	93	94	95	96	97	98	99

Adding Opposite Corners of Four Small Squares

Find any four small squares that meet in a single corner. Add the numbers in the two pairs of opposite corners. Compare sums.

REMEMBER:
Explore
Discover
Check
Describe
Explain

0	1	2	3	4	5	6	7	8	9
10	11	12	13	14	15	16	17	18	19
20	21	22	23	24	25	26	27	28	29
30	31	32	33	34	35	36	37	38	39
40	41	42	43	44	45	46	47	48	49
50	51	52	53	54	55	56	57	58	59
		63							

What happens with any size square?
What happens with any rectangle? Explain why.

6

Sum-Digit Pattern of Numbers on a Main Diagonal

Add the digits in the ones and the tens places of any numbers in any diagonal that runs from the upper right to the lower left. HINT: The digit in the tens place of numbers less than 10 is 0. Compare sums.

0	1	2	3	4	5	6	7	8	9
10	11	12	13	14	15	16	17	18	19
20	21	22	23	24	25	26	27	28	29
30	31	32	33	34	35	36	37	38	39
40	41	42	43	44	45	46	47	48	49
50	51	52	53	54	55	56	57	58	59
60	61	62	63	64	65	66	67	68	69
70	71	72	73	74	75	76	77	78	79
80	81	82	83	84	85	86	87	88	89
90	91	92	93	94	95	96	97	98	99

3

Sum-Digit Pattern of Consecutive Numbers in a Row

Add any six consecutive numbers in a row. EXAMPLE:

$$12 + 13 + 14 + 15 + 16 + 17 = 87$$

Add the two middle numbers.

EXAMPLE: $14 + 15 = 29$

If you multiply the sum of the two middle numbers by a certain number, you'll get the sum of the six numbers! Can you find that number?

Add the two outer numbers. What do you notice?

0	1	2	3	4	5	6	7	8	9
10	11	12	13	14	15	16	17	18	19
20	21	22	23	24	25	26	27	28	29
30	31	32	33	34	35	36	37	38	39
40	41	42	43	44	45	46	47	48	49

Challenge: Try the same thing with consecutive numbers in a column.

2

The Great X

Locate any square on the grid that is three numbers high by three numbers wide. Add the three numbers in the two diagonals between each of the corners. Compare the sums.

0	1	2	3	4	5	6	7	8	9
10	11	12	13	14	15	16	17	18	19
20	21	22	23	24	25	26	27	28	29
30	31	32	33	34	35	36	37	38	39
40	41	42	43	44	45	46	47	48	49
50	51	52	53	54	55	56	57	58	59

Challenge: What happens with any size square? Can you invent a magic trick?

7

Inners and Outers

Pick any four numbers that are an equal distance apart in one row of the number grid. Add the two outer numbers. Add the two inner numbers. Compare sums.

0	1	2	3	4	5	6	7	8	9
10	11	12	13	14	15	16	17	18	19
20	21	22	23	24	25	26	27	28	29
30	31	32	33	34	35	36	37	38	39
40	41	42	43	44	45	46	47	48	49
50	51	52	53	54	55	56	57	58	59

Magic Trick: Have a friend pick any four numbers an equal distance apart on the number grid and add the outer two numbers and the inner two numbers. Can you always guess the difference? Can you always guess the quotient?

4

Three in a Row

Pick any three numbers that are an equal distance apart in one row of the grid. Add the three numbers. Multiply the middle number by 3. Compare the sum and product.

0	1	2	3	4	5	6	7	8	9
10	11	12	13	14	15	16	17	18	19
20	21	22	23	24	25	26	27	28	29
30	31	32	33	34	35	36	37	38	39
40	41	42	43	44	45	46	47	48	49

Magic Trick: Have a friend pick three numbers an equal distance apart on the grid. Add the numbers. Multiply the middle number by 3. Guess the difference if your friend subtracts the sum from the product. Guess the quotient if he or she divides.

5

What's My Line?

How do you make a curved line out of a straight line?

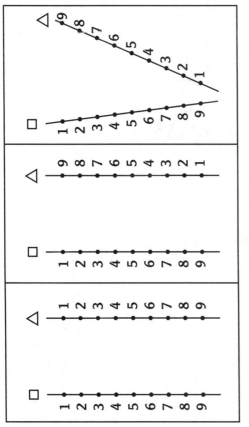

Open the book and find out!

Connect the numbers on the □ side of the angle to the numbers on the △ side of the angle. Use the table and equation.

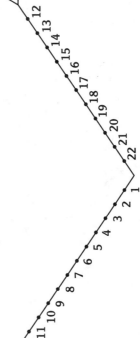

□	△
1	12
2	13
3	14
4	15
5	16
6	17
7	18
8	19
9	20
10	21
11	22
□ + 11 = △	

On each pair of lines below, connect 1 □ to 1 △, 2 □ to 2 △, 3 □ to 3 △, and so on.

Why do straight lines make curves?

Calculate to find the missing numbers in the table. Connect points on the zigzag line. Complete the table on a separate piece of paper. **HINT: Each number on the middle line will be connected to two other numbers.**

□	△
1	12
2	13
3	14
4	15
5	16

□	△
19	30
20	31
□ + 11 = △	

Calculate to find **all** the missing numbers in the table. Complete the table on a separate piece of paper. Connect the numbers around the perimeter of the triangle.

□	△
1	13
2	14
3	15
4	16
5	17
6	18
21	33
22	34
23	35
□ + 12 = △	

7

Fill in the missing numbers in the table. Connect the numbers on the □ line to the numbers on the △ line. Use the table and equation.

□	△
1	3
2	6
3	9
4	12
5	15
6	18
7	
8	
9	
10	
11	
12	36
13	39
□ × 3 = △	

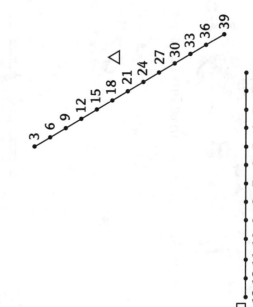

5

Use a straightedge. Draw line segments to connect the numbers on the □ side of the angle to the numbers on the △ side of the angle. The point connection table and equation show how to connect the points. The first one is done for you.

□	△
0	10
1	9
2	8
3	7
4	6
5	5
6	4
7	3
8	2
9	1
10	0
□ + △ = 10	

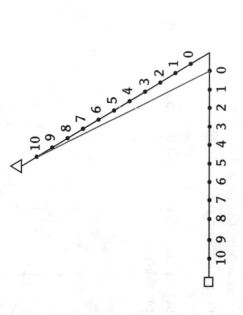

2

Connect the numbers on the □ side of the angle to the numbers on the △ side of the angle. Use the table and equation.

□	△
1	10
2	11
3	12
4	13
5	14
6	15
7	16
8	17
9	18
□ + 9 = △	

Color in spaces between the lines to make an interesting design.

4

Get in Shape!

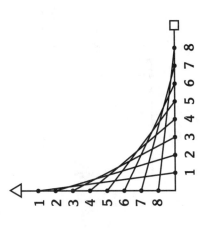

□	△
1	1
2	2
3	3
4	4
5	5
6	6
7	7
8	8
□ = △	

How do you get in shape? Exercise your mind!

Complete the point connection table on a separate piece of paper. Connect the numbers around the circle. The table and equation tell you how to connect the points. Add numbers around the circle as you need them.

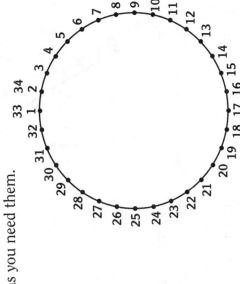

□	△
1	2
2	4
3	6
4	8
5	10
6	12

29	58
30	60
31	62
2 × □ = △	

22 Math Puzzle Mini-Books Scholastic Professional Books

Invent your own design on the hexagon below.

Invent your own design on the square below.

Draw line segments to connect the □ numbers to the △ numbers in each inscribed triangle in the hexagon. The table and equation tell you how to connect the points.

HINT: Each point will be connected to two other points.

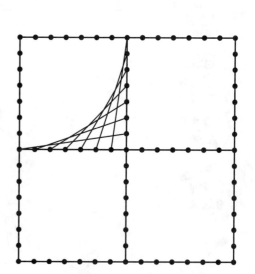

□	△
1	9
2	10
3	11
4	12
5	13
6	14
7	15
8	16
$\square + 8 = \triangle$	

Complete the point connection table on a separate sheet of paper. Draw line segments to connect the □ numbers to the △ numbers on the circle. Add numbers around the circle as you need them.

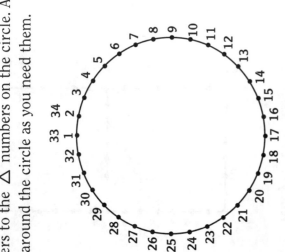

□	△
1	7
2	8
3	9
4	10
5	11
6	12

30	36
31	37
32	38
$\square + 6 = \triangle$	

22 Math Puzzle Mini-Books Scholastic Professional Books

Look at the pattern on the upper right quadrant of the square. Continue the same pattern on the other three quadrants.

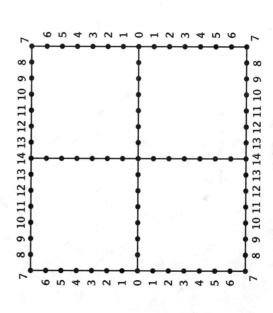

Connect the numbers on each small square according the table and equation.

□	△
0	8
1	9
2	10
3	11
4	12
5	13
6	14
$\square + 8 = \triangle$	

Four pockets have pennies.

The same number of pennies is in each pocket.

The sum of the number of pennies in all four pockets is greater than 16 and less than 24.

How many pennies are in each pocket?

8

Pockets Full of Pennies

In each of the following puzzles, someone has pennies in 1, 2, 3, or 4 pockets. It's up to you to guess how many pennies are in each pocket.

22 Math Puzzle Mini-Books Scholastic Professional Books

Two pockets contain pennies.

The right pocket has 2 more pennies than the left pocket.

The right pocket has 2 times as many pennies as the left pocket.

The number of pennies in the left pocket times itself equals the number of pennies in the right pocket:

Pennies in Left Pocket
x Pennies in Left Pocket
Pennies in Right Pocket

How many pennies are in each pocket?

6

Four pockets contain pennies.

Each pocket contains a different number of pennies.

The sum of all the pennies is less than 11.

How many pennies are in each pocket?

3

Four pockets have pennies.

Each pocket has a different number of pennies.

No pocket has more than 4 pennies.

How many pennies are in each pocket?

Lyle puts pennies in three pockets.

He puts a different number of pennies in each pocket.

The difference in the number of pennies between the pocket with the most pennies and the least pennies is 2.

The sum of the number of pennies in the three pockets is 12.

How many pennies are in each pocket?

22 Math Puzzle Mini-Books · Scholastic Professional Books

Four pockets have pennies.

The sum of the number of pennies in all four pockets is 5.

How many pennies are in each pocket?

There are pennies in two pockets.

The sum of the number of pennies in the two pockets is 10.

The difference in the number of pennies is 0.

How many pennies are in each pocket?

Four pockets have pennies.

There is a different number of pennies in each pocket.

The number of pennies in one pocket is equal to $24 \div 4$.

The number of pennies in another pocket is equal to $35 \div 5$.

The number of pennies in another pocket is equal to $48 \div 6$.

The sum of the number of pennies in all four pockets is 26.

How many pennies are in each pocket?

8

22 Math Puzzle Mini-Books Scholastic Professional Books

More Pockets, More Pennies

In each of the following puzzles, Pamela puts pennies in 1, 2, 3, or 4 pockets.

How many pennies are in each pocket?
Read the clues and solve the puzzles.

Two pockets contain pennies.

There is a different number of pennies in each pocket.

The number of pennies in each pocket is an even number.

The sum of the number of the pennies in both pockets is 14.

The pocket that contains the least number of pennies has more than 5 pennies.

How many pennies are in each pocket?

6

Pamela puts pennies in three pockets.

She slips a different number of pennies in each pocket.

The difference in the number of pennies in the pocket with the most pennies and the least pennies is 2.

The sum of the number of pennies in all three pockets is 15.

How many pennies are in each pocket?

3

Pamela puts pennies in two pockets.

She puts a different number of pennies in each pocket.

The difference in the number of pennies is 4.

The sum of the number of pennies is 10.

How many pennies are in each pocket?

2

Pamela's pants have four pockets.

Three pockets have the same number of pennies.

One pocket has one more penny than the other pockets.

The sum of the number of pennies in the four pockets is 21.

How many pennies are in each pocket?

7

Pamela puts pennies in four pockets.

There is a different number of pennies in each pocket.

The pocket with the most pennies contains fewer than 7 pennies.

The sum of the number of pennies in all four pockets is 18.

How many pennies are in each pocket?

4

There are pennies in one pocket of Pamela's coat.

The pocket has more than 5 pennies in it.

The number of pennies in the pocket times itself is less than 50.

The pocket has an odd number of pennies.

How many pennies are in the pocket?

5

22 Math Puzzle Mini-Books Scholastic Professional Books

Odds

You'll need: 2 players, 15 markers (beans, pebbles, buttons, or other small objects)

Rules:

Place all 15 markers between the players.

Players take turns picking up either one, two, or three markers at a time until all the markers are gone.

And the Winner Is!

The player who has an odd number of markers at the end of the game wins.

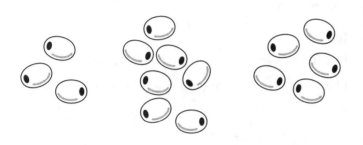

8

Nim Games for Two Players

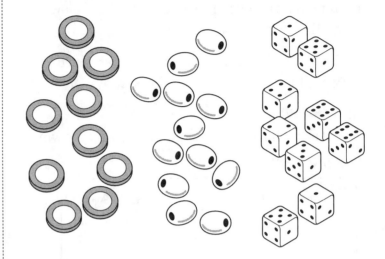

Play each game several times until you discover a successful winning strategy. Then move on to the next game.

22 Math Puzzle Mini-Books Scholastic Professional Books

Fill Up

You'll need: 2 players; colored pencils or crayons; 3 x 6 rectangle drawn on graph paper or game board

To make the game board:

Fold a sheet of paper lengthwise into thirds. Draw a line along each fold. Fold the paper in half with the short sides together. Then fold it into thirds. Draw a line along each fold. You should have a game board with six spaces across and three spaces down.

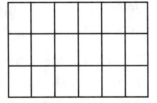

Rules:

Players take turns coloring in either one or two empty spaces on the game board.

The spaces do not have to be next to each other.

And the Winner Is!

The player who colors in the last space wins.

6

Dicey Count Up

You'll need: 2 players who can count from 1 to 32, a pair of dice

Rules:

Before beginning the game, each player rolls one die. The two numbers are added to 20 to get the target number (the number to play to). The player with the higher roll goes first.

Players take turns counting from 1 to the target number.

Each player may say one, two, or three numbers that follow each other. They must start counting where the other player ended.

And the Winner Is!

The player who says the target number wins the game.

3

Count Up

You'll need: 2 players who can count from 1 to 21

Rules:

Players take turns counting from 1 to 21.

Each player says one or two numbers that follow each other. EXAMPLE: 4, 5 but not 4, 8

Each player starts where the other player ended. EXAMPLE: Player 1 says, "5, 6." Player 2 says, "7, 8."

And the Winner Is!

The player who says 21 loses. The other player wins.

Sample Game:

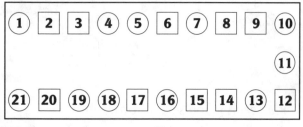

☐ **Player wins and** ◯ **Player loses**

22 Math Puzzle Mini-Books Scholastic Professional Books

Run the Track

You'll need: 2 players, a racetrack, 1 marker (a bean, pebble, or button) that will fit in the spaces of the racetrack

To make the racetrack:

On a sheet of paper, draw the squares of a racetrack along two long sides and one short side. You should have 35 squares. Label the squares at the top of the page **Start** and **Finish** like the racetrack shown here.

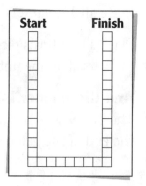

Rules:

Put the marker in the first square below **Start**.

Players take turns moving the marker 1, 2, 3, or 4 spaces along the track.

And the Winner Is!

The first player who moves onto **Finish** wins.

Pick Up

You'll need: 2 players, 16 game markers (pebbles, buttons, beans, or other small objects)

Rules:

Players put all 16 markers between them.

They take turns picking up either one, two, or three markers at a time.

And the Winner Is!

The player who picks up the last marker loses. The other player wins.

Calculator Countdown

You'll need: 2 players, a calculator

Rules:

One of the players enters the number 35 into the calculator.

Players take turns subtracting either 1, 2, or 3 from the number showing in the calculator display. EXAMPLE: 23 shows in the display. Player 1 subtracts 2 by pressing ⊟ ② ⊜. The display then reads 21.

And the Winner Is!

The player who gets 0 to show in calculator display.

Penny and Dime Magic

Say to a friend:
If you hide a penny in one hand and a dime in the other hand, I can tell you where each coin is hidden.

1. Hide a penny in one hand and a dime in the other hand—without letting me see in which hands the coins are hidden. EXAMPLE: left hand—penny, right hand—dime

2. Multiply the value of the coin in the left hand by 2, 4, 6, or 8. EXAMPLE: 1¢ × 6 = 6¢

3. Multiply the value of the coin in the right hand by 1, 3, 5, 7, or 9. EXAMPLE: 10¢ × 3 = 30¢

4. Add the two products together and tell me the sum: EXAMPLE: 6¢ + 30¢ = 36¢

5. The penny is in your left hand.

Here's the magic: If the sum is odd, the penny is in the right hand. If the sum is even, the penny is in the left hand. The dime is in the other hand.

8

22 Math Puzzle Mini-Books Scholastic Professional Books

Magical Math!

Try these magic tricks on friends. Follow the steps below and—who knows?—you may become a great magician!

☆ Practice a trick several times before performing it for someone. This will also help you discover how the trick works.

☆ Examples for the tricks appear in the book.

☆ If you need a calculator or paper and pencil for a trick, have them ready to use.

☆ Do not do the same trick for someone over and over again—or else he or she'll figure out how it works!

Even I Win, Odd You Lose

Magician (that's you): Here's an amazing magic trick. You pick a number. DON'T TELL ME WHAT IT IS. I'll ask you to do some calculations. If the answer is even, I win. If the answer is odd, you lose.

Friend: Hey! That's not fair!

Magician: OK. Even I win. Odd you win.

1. Pick a number. **21**
2. Add 4 to it. . . . . **21 + 4 = 25**
3. Multiply the sum by 4. . . . . **25 x 4 = 100**
4. Subtract 4 from the product. . . . . **100 − 4 = 96**
5. Divide by 2.
   What is the quotient? . . . . **96 ÷ 2 = 48**

Here's the magic: The answer is even—the answer will *always* by even, so you *always* win.

6

Two-Digit Number Magic

Say this to friend:
Pick a 2-digit counting number and a 1-digit counting number (not 0). DO NOT tell me what they are. I'll ask you to do some calculations. Then I'll tell you what your numbers are!

1. Write down your two numbers. **35, 7**
2. Multiply the 2-digit number by 10. . . . . **35 x 10 = 350**
3. Add the 1-digit number to the product. . . . . **350 + 7 = 357**
4. Tell me the sum. **357**

Here's the magic: Your friend's original 2-digit number appears in the hundreds and tens places of the sum, and the 1-digit number is the number in the ones place of the sum!

3

Two Number Magic

Say this to a friend:
Pick two 1-digit counting numbers (not 0).
DO NOT show them to me. I'll ask you to do
some calculations. Then I'll tell you what your
numbers are!

1. Write down your two numbers. 3, 2

2. Multiply the first number by 10. . . . 3 x 10 = 30

3. Add the second number
   to the product. . . . . 30 + 2 = 32

4. Tell me the sum. 32

Here's the magic: Your friend's original two numbers are the digits in the tens and ones places of the sum!

After you figure out how this trick works, can you do the trick with three numbers? Multiply the first number by 100, the second number by 10, and the third number by 1. Then add the three products.

2

22 Math Puzzle Mini-Books Scholastic Professional Books

Penny, Nickel, and Dime Magic

1. Hide a penny, nickel, and dime in one of your hands. Don't tell what you have hidden.

2. Find a friend who has a penny and say:

   Guess what will happen when you add your penny to my money. Here's what can happen: If you guess even and the total amount is even, you win. If you guess odd and the total amount is odd, you win. Otherwise, I win.

3. Have your friend guess odd or even.

4. If your friend guesses even, here's what you do: Put his or her penny with your penny, nickel, and dime, and add the amount of money. The sum is 17¢, which is an odd number, so you win!

   If your friend guesses odd: Put his or her penny with your penny, nickel, and dime, and count the number of coins. You have four coins, which is an even number, so you win! Magic!

5. Give back the penny to your friend.

7

Two Number Magic #2

Say this to a friend:
Pick two 1-digit counting numbers (not 0).
DO NOT show them to me. I'll ask you to do
some calculations. Then I'll tell you what your
numbers are!

1. Write down your two numbers. . . . 3, 5

2. Multiply the first number by 5. . . . 3 x 5 = 15

3. Add 2 to the product. . . . . 15 + 2 = 17

4. Multiply the sum by 2. . . . . 17 x 2 = 34

5. Subtract 4 from the product. . . . . 34 – 4 = 30

6. Add the second number
   to the difference. . . . . 30 + 5 = 35

7. Tell me the sum. 35

Here's the magic: Your friend's original two numbers appear in the tens and ones places of the sum!

4

Even I Win, Odd I Win

Say this to a friend:
We each secretly write down a counting number
(not 0). I'll tell you if the sum of our numbers is
even or odd—without looking at your number!

1. Your friend writes down a number. 12

2. You always write down any **odd** number. . . . 5

3. **Say:** If your number is even and we add our numbers, the sum will be an odd number. If your number is odd, and we add our numbers, the sum will be an even number.

4. Show your numbers to each
   other and then add them. . . . . 12 + 5 = 17

Here's the magic: Your friend picked an even number. You picked an odd number. The sum is odd.

The trick is that you must always pick an odd number:

even + odd = odd and **odd + odd = even**

5

Speed Add With Two Friends

Find two friends who will to try to add faster than you.

1. Have the first friend write down a counting number (not 0).

2. Ask the second friend write down a larger counting number under the first number.

3. Tell the first friend to add the two numbers and write the sum underneath the second number.

4. Instruct your friends to take turns adding the last two numbers in the list and writing the sum at the bottom of the list. They repeat this until there are a total of 10 numbers in the list.

Have your friends show the list to you. Challenge them to add the column of numbers faster than you. They can even use a calculator.

Here's the magic: The sum is 11 times the seventh number on the list! **111 x 11 = 1,221**

8

3
12
15
27
42
69
111
180
291
471

22 Math Puzzle Mini-Books Scholastic Professional Books

MORE Magical Math!

Amaze your friends with these magic tricks.

☆ Practice, practice, practice! The book shows examples for the tricks.

☆ Have paper and pencils and a calculator ready.

☆ Try these things to discover how the tricks work: Use different numbers each time you do the trick. Explain to a friend why you think the trick works.

This Trick Is About 1

Say this to a friend:

1. Secretly write down a counting number. . . . **5**

2. Add 2 to the number. . . . . **5 + 2 = 7**

3. Multiply the sum by 2. . . . . **7 x 2 = 14**

4. Add 2 to the product. . . . . **14 + 2 = 16**

5. Multiply the sum by 2. . . . . **16 x 2 = 32**

6. Divide the product by 2. . . . . **32 ÷ 2 = 16**

7. Subtract 2 from the quotient. . . . . **16 – 2 = 14**

8. Divide the difference by 2. . . . . . . **14 ÷ 2 = 7**

9. Subtract your original number. . . . **7 – 5 = 2**

10. Divide the difference by 2. . . . . . . **2 ÷ 2 = 1**

Here's the magic: This trick is about the number 1. The answer is *always* 1!

Can you change the trick so that a friend adds, subtracts, multiplies, and divides by a number other than 2—and still ends up with an answer of 1?

6

Double Reverse

1. Ask a friend to write down any 3-digit number that doesn't contain any zeros. Remind her or him not to show the number to you.

2. Tell your friend to reverse the order of the digits and write down the new number.

3. Now your friend subtracts the smaller number from the larger number.

4. Then your friend reverses the order of the digits in the difference. He or she writes that new number below.

5 He or she adds the numbers.

①	175
②	571
③	571
	– 175
	396
④	693
⑤	396
	+ 693
	1,089

Here's the magic: Without seeing any of the calculations, you know that the sum is 1,089! Use magic to check the subtraction.

Say: Now add the first and last digits of the difference. They will equal the middle digit. Magic! In this example, 3 + 6 = 9.

3

Descend and Reverse

1. Write the number **198** on a piece of paper. Hide it in your pocket.

2. **Say to a friend:** Pick a 3-digit number. DO NOT tell me what it is. I'll ask you to do some calculations. Then I'll tell you the number. OK, write down a 3-digit number that has no zeros. Its digits should decrease by 1 from left to right.

3. Reverse the order of the digits, and write down the new number below the original number.

4. Subtract the smaller number from the larger number.

5. Slap your pocket two times. Pull out the piece of paper with the answer 198 on it. Amazing!

$$
\begin{array}{r}
543 \\
345 \\
\hline
543 \\
-345 \\
\hline
198
\end{array}
$$

You can do the same trick with a 4-digit number. The answer will always be 3,087.

22 Math Puzzle Mini-Books Scholastic Professional Books

More Two Number Magic

Challenge a friend to do the following:

1. Secretly write down two 1-digit numbers (not 0). **2, 8**

2. Add 2 to the first number. . . . . . . **2 + 2 = 4**

3. Multiply the sum by 5. . . . . **4 x 5 = 20**

4. Add 2 to the product. . . . . **20 + 2 = 22**

5. Multiply the sum by 2. . . . . **22 x 2 = 44**

6. Subtract 4 from the product. . . . . **44 − 4 = 40**

7. Add the second number to the result. . . . . **40 + 8 = 48**

8. Subtract 20 from the sum. . . . . . . **48 − 20 = 28**

9. Ask your friend to tell you the result. **28**

Here's the magic: Your friend's original numbers are the digits in the tens and the ones places of the final difference! Hmm—where have you seen this magic trick before?

Magic Sub 5

Say this to a friend:

1. Secretly write down a number— remember, no zeros! **8**

2. Add 4 to it. . . . . **8 + 4 = 12**

3. Multiply the sum by 4. . . . . **12 x 4 = 48**

4. Add 4 to the product. . . . . **48 + 4 = 52**

5. Divide the sum by 4. . . . . **52 ÷ 4 = 13**

6. Tell me the answer. **13**

Here's the magic: To find your friend's original number, subtract 5 from the answer he or she gives you!

A Fantastic Two-Person Trick

1. Write the number **12** on a piece of paper. Put it in your pocket.

2. Have Person 1 pick a number between 10 and 20 and record it.

Person 1
15

3. Have Person 2 double Person 1's number and record it.

Person 2
15 + 15 = 30

4. Subtract 4 from the number Person 1 has. Add 4 to the number Person 2 has.

Person 1
15 − 4 = 11

Person 2
30 + 4 = 34

5. Multiply Person 1's difference by 2. Add the product to Person 1's difference. Subtract the product from Person 2's number.

Person 1
11 x 2 = 22
11 + 22 = 33

Person 2
34 − 22 = 12

Here's the magic: Slap your pocket two times. Remove the piece of paper with the number 12— Person 2's final number!

Write each of the numbers **1**, **2**, **3**, **4**, **5**, and **6** once in the circles so that the three numbers in each line add up to **12**.

8

Abracadabra!

See if you can solve the "magic" puzzles in this book.

Write each of the numbers **0**, **1**, **2**, **3**, and **4** once in the circles so that the sum of the three numbers in each straight line is **6**.

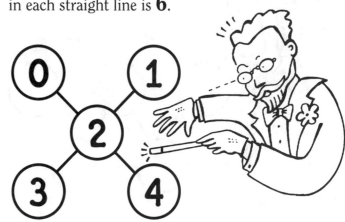

22 Math Puzzle Mini-Books Scholastic Professional Books

Write each of the numbers **1**, **2**, **3**, **4**, **5**, and **6** once in the circles so that the three numbers in each line add up to **10**.

6

Write each of the numbers **1**, **3**, **5**, **7**, **9**, **11**, and **13** once in the circles so that the three numbers in each line add up to **21**.

3

Write each of the numbers **1**, **2**, **3**, **4**, **5**, **6**, and **7** once in the circles so that the three numbers in each line add up to **12**.

Write each of the numbers **1**, **2**, **3**, **4**, **5**, and **6** once in the circles so that the three numbers in each line add up to **11**.

22 Math Puzzle Mini-Books Scholastic Professional Books

2

7

Write each of the numbers **2**, **4**, **6**, **8**, **10**, **12**, and **14** once in the circles so that the three numbers in each line add up to **24**.

Write each of the numbers **1**, **2**, **3**, **4**, **5**, and **6** once in the circles so that the three numbers in each line add up to **9**.

4

5

HOT

You'll need: 2 players, 9 small index cards with the following words printed on them: *hot, hear, tied, form, wasp, brim, tank, ship, woes*

Rules:

Place the cards faceup between the players.

Players take turns.

During a turn, a player draws a card and places it faceup in front of himself or herself.

And the Winner Is!

The player to draw three cards that contain the same letter, for example three words that each have an *s* wins. Players can have more than three cards before the game ends.

A Tie Game

If neither player has three words with the same letter when all the cards have been drawn, the game ends in a tie.

8

22 Math Puzzle Mini-Books Scholastic Professional Books

Name_____

Zim! Zam! Zoom!

See if you can solve the "magic" puzzles in this book.

Write each of the numbers **0**, **1**, **2**, **3**, and **4** once in the squares of the figure so that the three numbers in the two diagonals add up to **6**.

3

Write each of the numbers **1**, **2**, **3**, **4**, **5**, **6**, **7**, **8**, and **9** once in the squares so that the sums of the three numbers in the rows across, the columns down, and the main diagonals are equal to **15**.

6

Write each of the numbers **2**, **4**, **6**, **8**, and **10** once in the squares of the figure so that the three numbers in the row across and the column down are equal to **18**.

Write each of the numbers **1**, **2**, **3**, **4**, and **5** once in the squares of the figure so that the three numbers in the row across and the column down are equal to **9**.

2

Get 15

You'll need: 2 players, the numbers 1–9 written on a piece of paper, 2 different colored pencils or markers

Rules:

Each players uses a different color pencil or marker.

Players take turns.

During a turn, a player circles one of the numbers from 1–9. A number can be circled only once.

And the Winner Is!

The first player to circle three numbers that add up to 15 wins. A player may circle more than three numbers, but three of the numbers *must* add up to 15.

Sample Game

1 ②③4⑤6⑦8 9

Move 1: ◯ takes 7. ☐ takes 8.
Move 2: ◯ takes 6. ☐ takes 2.
Move 2: ◯ takes 5. ☐ takes 4.
Move 4: ◯ takes 3—and wins!
3 + 5 + 7 = 15

7

Write each of the numbers **1**, **2**, **3**, **4**, **5**, **6**, **7**, and **8** once in the squares so that the sums of the three numbers in the rows across and the columns down are equal to **12**.

4

Write each of the numbers **1**, **2**, **3**, **4**, **5**, **6**, and **7** once in the squares so that the sums of the three numbers in the rows across, the columns down, and the main diagonals are equal to **12**.

5

Guess what? The number sentence below is false.
Can you move one crayon to make it true?

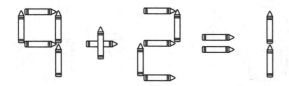

Solution:

22 Math Puzzle Mini-Books Scholastic Professional Books

Crazy Crayon
Digits

0123456789

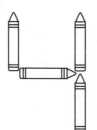

Can you change the 4 into a 9? Can you do it by adding just one crayon?

Solution: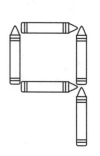

Are you ready to create some crazy crayon digits?
Then grab your crayons and turn the page.

The number sentence below is false.
Can you move one crayon to make it true?

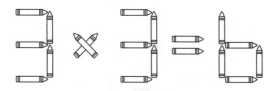

Draw your solution below.

Use crayons to make a 5. Then move
one crayon to change the 5 into a 3.

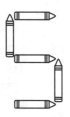

Draw your solution below.

Use crayons to make a 9.
Then move one crayon to change the 9 into a 5.

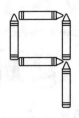

Draw your solution below.

The number sentence below is false.
Can you move one crayon to make it true?

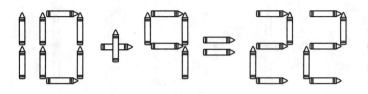

Solution:

Make a 3. Then move one crayon
to change the 3 into a 2.

Draw your solution below.

Make a 5. Then move one crayon
to change the 5 into a 6.

Draw your solution below.

22 Math Puzzle Mini-Books Scholastic Professional Books

Use 24 crayons to make this construction.

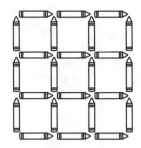

Remove four crayons and leave five equal squares.

Solution:

22 Math Puzzle Mini-Books Scholastic Professional Books

Creative Crayon Constructions

In each of these puzzles, you'll rearrange, add, or remove crayons to make new shapes.

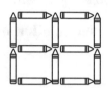

Remove two crayons and leave exactly three squares.

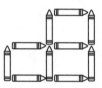

You can also use toothpicks, straws, or coffee stirrers. Oh, by the way—there may be more than one way to solve a puzzle.

Can you make this fish swim in the opposite direction? Can you do it by moving only three crayons?

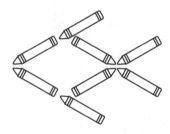

Draw your solution below.

Make this crayon creature.

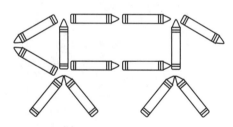

Turn around the creature. Move one crayon.

Draw your solution below.

Use crayons to make this house.

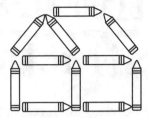

Make two houses by moving one crayon.

Draw your solution below.

Use 16 crayons to make this construction.

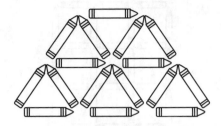

Move four crayons and turn
the construction upside down.

Solution:

22 Math Puzzle Mini-Books Scholastic Professional Books

Change these two rectangles into two squares.

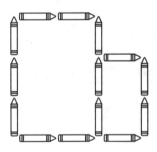

Move only two crayons.

Draw your solution below.

Create this goblet.

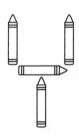

Turn it upside down. Move only two crayons.

Draw your solution below.

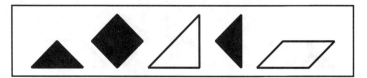

Use the three shaded tans to make a large rectangle. What other shape did you make out of these same three tans?

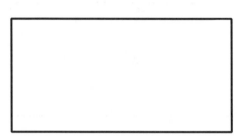

Use the same tans to make a trapezoid and a parallelogram. Trace your results on another piece of paper.

8

22 Math Puzzle Mini-Books Scholastic Professional Books

Name_____

▽TRIANGLE▽TROUBLE!

Tan accidentally dropped his triangle-shaped tile. It broke into 5 pieces—2 small triangles, 1 medium triangle, 1 square, and 1 parallelogram. This is how he put the triangle back together.

Before   **After**

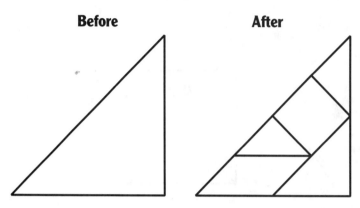

Tan named the pieces *tans*. He found out that he could put the tans together to make different shapes. Cut out and use your tans.

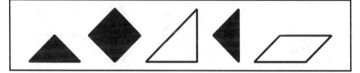

Make a large triangle. Use the three shaded pieces.

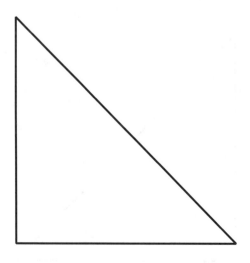

6

Tan made a parallelogram out of the two shaded pieces. Can you make the shape? Draw lines to show your solution.

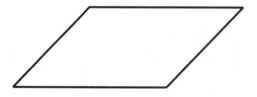

Now use the same two shaded shapes to make a triangle.

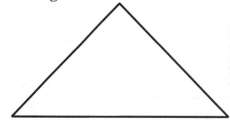

3

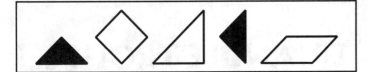

Tan made a square out of the two shaded pieces. Can you figure out how he did it? Draw lines to show your solution. (This one's done for you.)

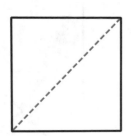

2

Use the three shaded tans to make a rectangle.

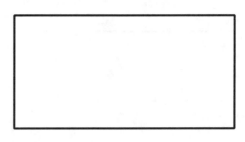

Now use the same three tans to make a square and a parallelogram. Trace your results on another piece of paper.

7

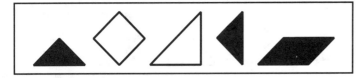

Make a large triangle using the three shaded pieces. Draw lines to show your solution.

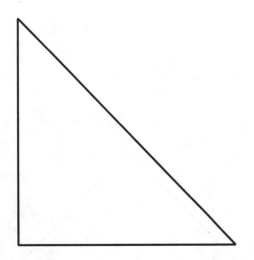

4

Tan made another large triangle. He used the three shaded pieces. Can you figure out how he did it? Draw lines to show your solution.

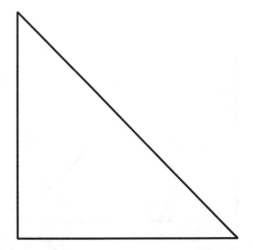

5

22 Math Puzzle Mini-Books Scholastic Professional Books

What kind of skunk doesn't smell? A tangram skunk! Construct the skunk. Show your solution.

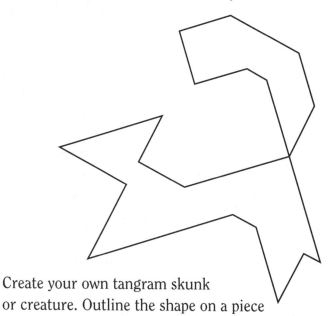

Create your own tangram skunk or creature. Outline the shape on a piece of paper. Then give it to a friend to solve.

8

Name_____

THE MAGIC TILE

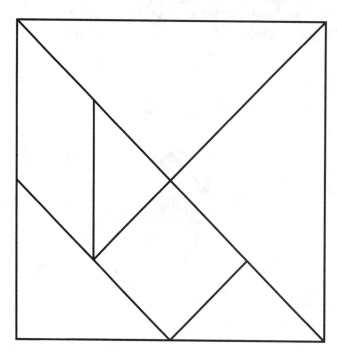

This is a tangram. What's magic about its pieces? Turn the page and find out.

22 Math Puzzle Mini-Books Scholastic Professional Books

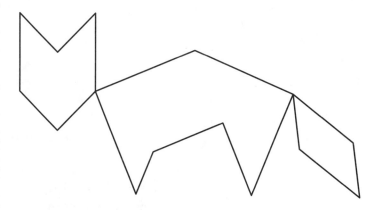

Construct the fox. Show your solution.

Create your own tangram fox or creature. Outline the shape on a piece of paper. Then give it to a friend to solve.

6

Use your tangram pieces to construct the lion. Then draw lines below to show your solution.

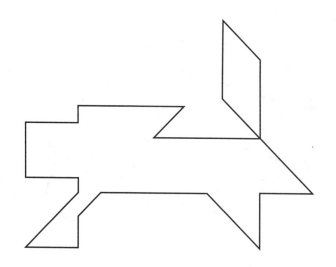

3

Use your tangram pieces to construct the owl.
Then draw lines below to show your solution.

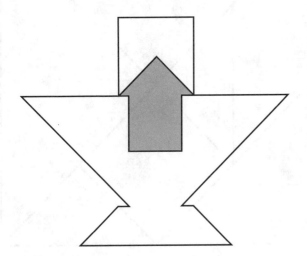

22 Math Puzzle Mini-Books Scholastic Professional Books

Construct the swan. Show your solution.

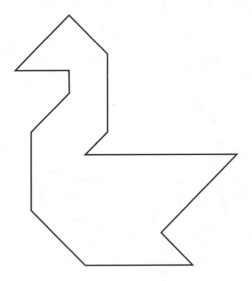

Create your own tangram swan or another bird.
Outline the shape on a piece of paper.
Then give it to a friend to solve.

Use your tangram pieces to construct the bear.
Then draw lines below to show your solution.

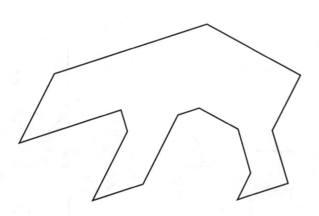

Construct the rabbit. Show your solution.

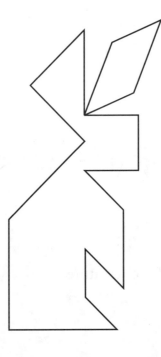

Create your own
tangram rabbit or
creature. Outline the
shape on a piece of
paper. Then give it to
a friend to solve.

Coin Cover-Up

Put your finger on any circle. From there, count 1, 2, 3. Place a penny on the circle that is 3. Continue doing this until all the circles have pennies—except one. Each time you must start on an empty circle.

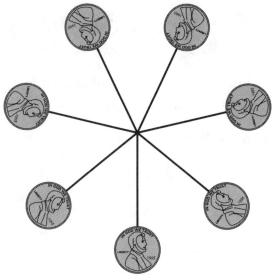

Can you figure out how to solve the Coin Cover-Up no matter which circle you start from?

22 Math Puzzle Mini-Books Scholastic Professional Books

Coin Challenges

Here's your first challenge. Can you arrange 6 coins in 2 rows? You think that's easy?

Think again—each row must contain 4 coins.

See? The middle of each row has 2 coins stacked on top of each other.

row 2↓

row 1→

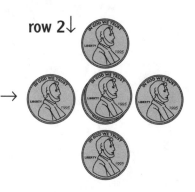

Odd Coins

Take 12 coins. Arrange them in a pattern that contains 3 straight lines with an odd number of coins in each line.

Draw your solution.

Coin Line Up

Use the line below. Place 3 coins so that there are two heads on one side of the line and two tails on the other.

_____

Draw or write your solution.

Coin Circle

Arrange 6 coins as shown in the drawing. Rearrange the coins into a circle by moving 2 coins. Which coins did you move?

Draw a picture to show how you solved the puzzle.

22 Math Puzzle Mini-Books Scholastic Professional Books

Coin Jump Game

Object: Make 4 stacks of penny pairs in 4 moves.

How to Play:
Arrange 8 pennies as shown below.

Pick up 1 penny at a time. Jump over 2 other pennies until you have 4 stacks of penny pairs.

Draw your solution.

Coin Pyramid

Arrange 10 coins in a pyramid as shown. Move just 3 coins to turn the pyramid upside down.

Draw your solution.

Puzzling H

Arrange 7 coins to form an H pattern. Counting diagonal, horizontal, and vertical lines, there are 5 rows with 3 coins in each row. Add 2 coins to create a new pattern that has 10 rows with 3 coins in each row. **Draw the coins below to show where you would place them.**

These are **Polypods**.

They share three characteristics.

None of these are Polypods.

Draw a Polypod. How do you know it's a Polypod?

8

Mystery Monsters

Can you spot the monster imposters? On each page, figure out the characteristics that each monster group shares. Then find the monsters that belong.

EXAMPLE: These are **Bipeddlers**.

They share one characteristic.

These are *not* Bipeddlers.

 Circle all the Bipeddlers.

2 feet

2 feet

22 Math Puzzle Mini-Books Scholastic Professional Books

These are **Hatterworts**.

They share two characteristics.

None of these are Hatterworts.

Draw a Hatterwort. How do you know it's a Hatterwort?

6

These are **Biojos**.

They share one characteristic.

These are *not* Biojos.

Circle all the Biojos.

3

These are **Liats**.
They share one characteristic.

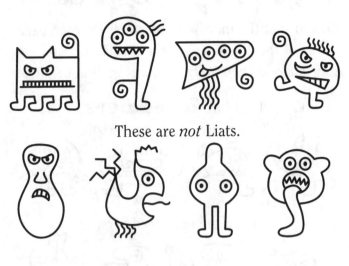

These are *not* Liats.

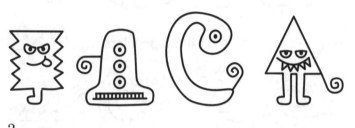

Circle all the Liats.

These are **Rhinobeaks**.
They share three characteristics.

None of these are Rhinobeaks.

Draw a Rhinobeak. How do you know it's a Rhinobeak?

22 Math Puzzle Mini-Books Scholastic Professional Books

These are **Taileroos**.
They share two characteristics.

These are *not* Taileroos.

Draw a Taileroo. How do you know it's a Taileroo?

These are **Snailies**.
They share two characteristics.

None of these are Snailies.

Draw a Snailie. How do you know it's a Snailie?

Hundreds Charts

Use with Great Grid Games and More Grid Games.

0	1	2	3	4	5	6	7	8	9
10	11	12	13	14	15	16	17	18	19
20	21	22	23	24	25	26	27	28	29
30	31	32	33	34	35	36	37	38	39
40	41	42	43	44	45	46	47	48	49
50	51	52	53	54	55	56	57	58	59
60	61	62	63	64	65	66	67	68	69
70	71	72	73	74	75	76	77	78	79
80	81	82	83	84	85	86	87	88	89
90	91	92	93	94	95	96	97	98	99

0	1	2	3	4	5	6	7	8	9
10	11	12	13	14	15	16	17	18	19
20	21	22	23	24	25	26	27	28	29
30	31	32	33	34	35	36	37	38	39
40	41	42	43	44	45	46	47	48	49
50	51	52	53	54	55	56	57	58	59
60	61	62	63	64	65	66	67	68	69
70	71	72	73	74	75	76	77	78	79
80	81	82	83	84	85	86	87	88	89
90	91	92	93	94	95	96	97	98	99

Tangrams

Use with Triangle Trouble! and The Magic Tile.

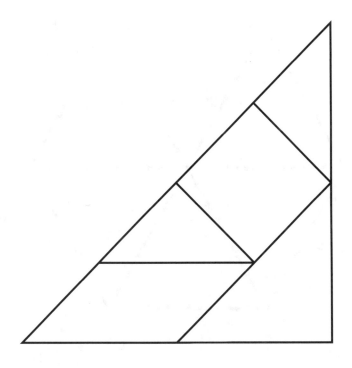

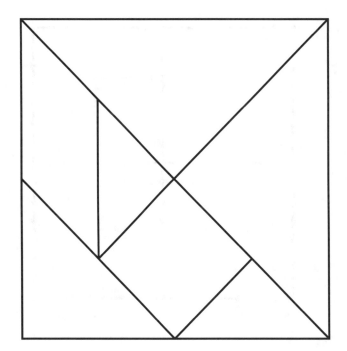

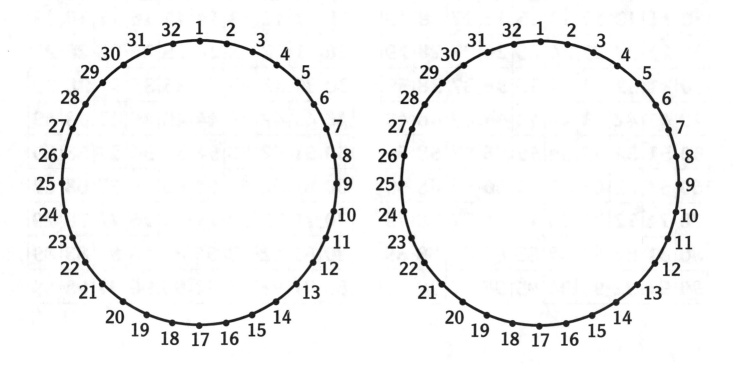

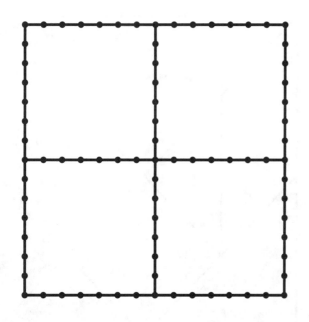

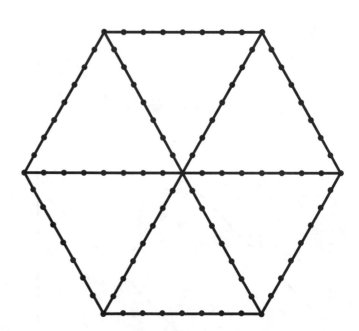

Answers

I'll Cross That Bridge When I Get to It, pp. 13-14

Multiple solutions exist. Each landmass is labeled E (even) or O (odd) based on the number of bridges it contains. To solve problems, students must either start or end on a landmass labeled O. If a problem contains more than two landmasses labeled O, it is impossible to solve.

p. 2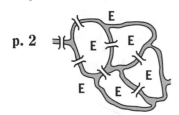

p. 3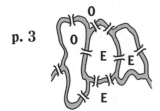

p. 4

p. 5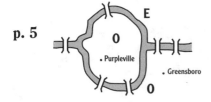

p. 6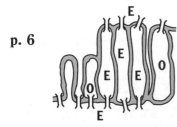

p. 7

MORE Bridge Crossing Problems, pp. 15-16

If a problem has 3 or more landmasses, each with an odd number of bridges, then it is impossible to complete the problem.

p. 2 school and park located on separate landmasses labeled O

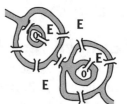

p. 3 houses located on separate landmasses labeled O

p. 4 impossible

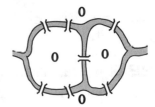

p. 5 impossible

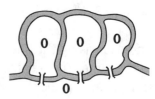

p. 6 store located on any landmass labeled O

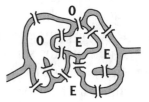

p. 7 impossible

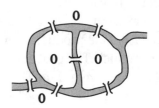

Don't Lift That Pencil!, pp. 17-18

Multiple solutions exist. The puzzle on page 5 is impossible to solve. Any puzzle that contains more than two odd vertices (vertices where an odd number of lines meet) is impossible to solve.

MORE Tracing Puzzles, pp. 19-20

Multiple solutions exist. The puzzles on pages 4, 6, 7, and 8 are impossible to solve. Each contains more than two odd vertices.

Don't Believe Your Eyes!, pp. 21-22

p. 1: There is no lowest step. **p. 2:** Both line segments are the same length. **p. 3:** Both box tops are the same size. **p. 4:** Yes, the sides are straight. **p. 5:** The lines are the same length. **p. 6:** line B **p. 7:** The circles are the same size. **p. 8:** The figures are the same size.

Great Grid Games, pp. 23-24

p. 2: 53, 44, 88, 36, 97, 65 **p. 3:** 15, 13, 45, 31, 35, 54 **p. 4:** 55, 24, 41, 97, 66, 55 **p. 5:** 96, 25, 45, 67 **p. 6:** −10, +10, +11, −11, −9, +9 **p. 7:** B = +12, C = +22, D = +18, E = +3, F = +20 **p. 8:** Multiple solutions exist.

MORE Grid Games, pp. 25-26

p. 2: Three times the sum of the inner two numbers or the outer two numbers equals the sum of all six numbers. **p. 3:** The sum of the digits is always the same because you are adding 9 by increasing the tens by 1 and decreasing the ones by 1. **p. 4:** The sums are always the same. This relates to finding averages. **p. 5:** The sum always equals the product. The equation for relating the sum and product is $a + (a + 1) + (a + 2) = 3a + 3 = 3(a + 1)$, where a is the number in the left corner. **p. 6:** The sums equal each other. The sums on one diagonal equal the sums on the other diagonal: $[a + (a + 11)] = (a + 1) + (a + 10) = 2a + 11$, where a is the number in the upper left corner. **p. 7:** The sums are equal to each other. The sums are the same as on page 6 except that you add the central number to both sides. **p. 8:** One product is always 10 more than the other product. The product of one diagonal is $a \times (a + 11) = a^2 + 11a$. The product of the other diagonal is $(a + 10)(a + 1) = a^2 + 11a + 10$. Here a is the number in the upper left corner.

What's My Line?, pp. 27-28

p. 2
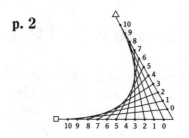

p. 3
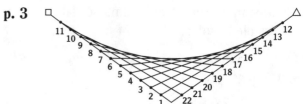

p. 4
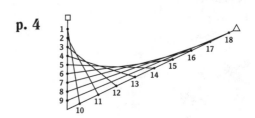

p. 5
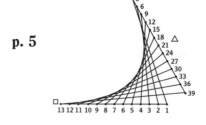

p. 6
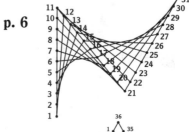

p. 7
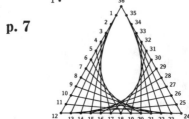

p. 8
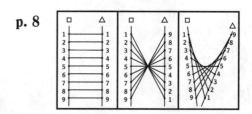

Get in Shape!, pp. 29-30

p. 2
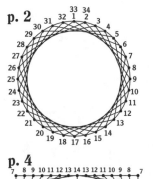

p. 3
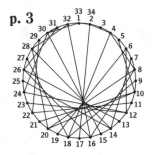

p. 4
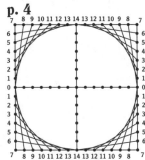

p. 5
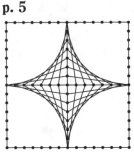

p. 7
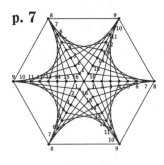

pp. 6 and 8 Check students' work.

Pockets Full of Pennies, pp. 31-32

p. 2: 1, 2, 3, 4 **p. 3:** 1, 2, 3, 4 **p. 4:** 1, 1, 1, 2 **p. 5:** 5, 5
p. 6: 2, 4 **p. 7:** 3, 4, 5 **p. 8:** 5, 5, 5, 5

MORE Pockets, More Pennies, pp. 33-34

p. 2: 7, 3 **p. 3:** 4, 5, 6 **p. 4:** 3, 4, 5, 6 **p. 5:** 7 **p. 6:** 6, 8
p. 7: 5, 5, 5, 6 **p. 8:** 5, 6, 7, 8

Nim Games, pp. 35-36

p. 2: winning numbers—2, 5, 8, 11, 14, 17, 20 **p. 3:** Winning numbers vary with throw of dice. **p. 4:** winning numbers—3, 7, 11, 15 **p. 5:** winning numbers—0, 4, 8, 12, 16, 20, 24, 28, 32 **p. 6:** The player who colors in the third, sixth, ninth, twelfth, fifteenth, and eighteenth squares wins. **p. 7:** The player whose marker lands on positions 5, 10, 15, 20, 25, 30, 35 wins.

Magic Circles, pp. 41-42

Answers may vary. Sample answers given.

p. 2
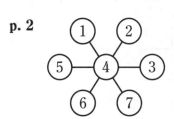

p. 3
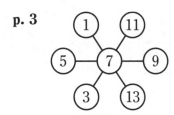

p. 4
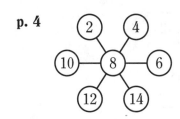

p. 5
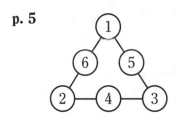

p. 6
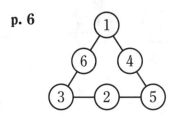

p. 7
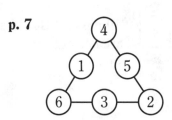

p. 8
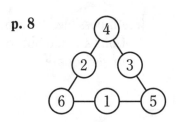

61

Magic Squares, pp. 43-44

Answers may vary. Sample answers given.

p. 2

	1	
2	3	4
	5	

p. 3

	2	
4	6	8
	10	

p. 4

3	7	2
8	■	4
1	5	6

p. 5

2	3	7
	4	
1	5	6

p. 6

2	9	4
7	5	3
6	1	8

p. 7

2	9	4
7	5	3
6	1	8

All possible Get 15 sums are laid out in this magic square.

p. 8

hot	form	woes
tank	hear	wasp
tied	brim	ship

All possible Hot combinations are laid out in this magic square.

Crazy Crayon Digits, pp. 45-46

p. 2

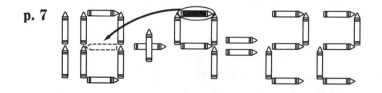

p. 3

p. 4

p. 5

p. 6

p. 7

p. 8

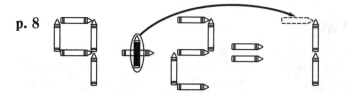

Creative Crayon Constructions, pp. 47-48

p. 2

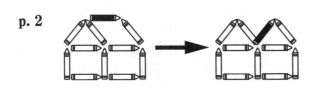

p. 3

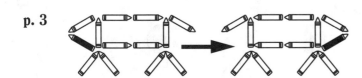

p. 4

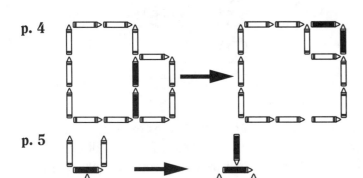

p. 5

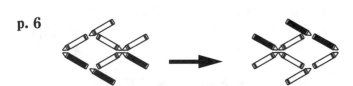

p. 6

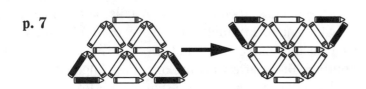

p. 7

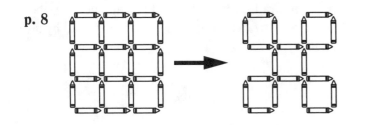

Triangle Trouble!, pp. 49-50

p. 3

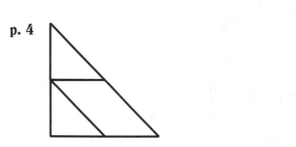

p. 4

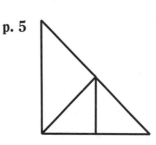

p. 5

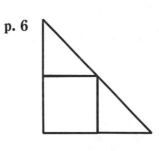

p. 6

p. 7

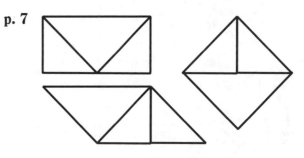

p. 8

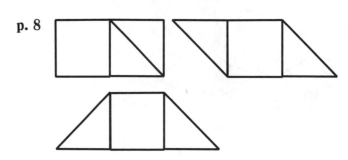

The Magic Tile, pp. 51-52

p. 2

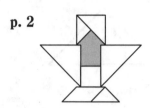

p. 3

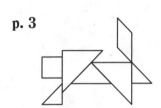

p. 4

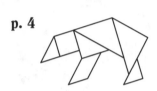

p. 5

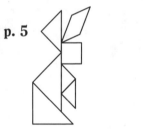

p. 6

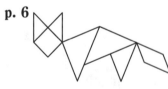

p. 7

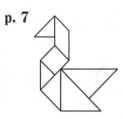

p. 8

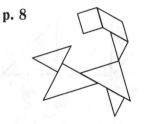

Coin Challenges, pp. 53-54

Possible solutions: **p. 2:** Move coin 1 so that it touches coins 4 and 5. Slide coin 5 in between coins 1 and 6. **p. 3:** Place one coin heads up on one side of line. Place a second coin tails up on the other side of the line. Stand a third coin on the line with its head facing the heads-up coin and its tail facing the tails-up coin. **p. 4:** Move coin 1 below coins 8 and 9. Move coin 10 beside coin 3. Move coin 7 beside coin 2. **p. 5:**

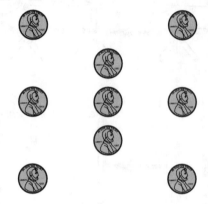

p. 6: Form a pyramid with 5 coins on each side. **p. 7:** Move coin 4 on top of coin 7. Move coin 6 on top of coin 2. Move coin 1 on top of coin 3. Move coin 8 on top of coin 5. **p. 8:** Start in any circle. Put your finger on that circle and count three from there, clockwise. Move to the first empty circle counterclockwise. Count three counterclockwise. Then continue counting three in any direction that places you in an empty circle.

Mystery Monsters, pp. 55-56

p. 1: A Bipeddler has two feet. **p. 2:** A Liat has a tail. **p. 3:** A Biojo has a two eyes. **p. 4:** A Taileroo has one curly tail and one triangle-shaped eye. **p. 5:** A Snailie's body has only curved lines and one oval-shaped mouth. **p. 6:** A Hatterwort has one three-fingered hand and one curved, concave side. **p. 7:** A Rhinobeak has one bowtie shape inside its body, one sharp beak, and one curly tail. **p. 8:** A Polypod has an even number of sides, two patterns on its body, and is bisected by a diagonal.